THE BEAM OF LIGHT MOVING THROUGH THE
MISTY CLOUD REACHED ZERO MILES PER HOUR.
ITS INFORMATION FLOW HAD CEASED
COMPLETELY. TIME HAD BEEN STOPPED.

Stop time and it is no longer impossible to imagine traveling through the frozen temporal landscape. Will past and
future spread all around us like a foreign country waiting to
be explored? Could we dream of finding the right vehicle
and a road map that would plot a course into its midst—
perhaps a course taking us into the past or the future?

A ray of light had been stopped dead in a laboratory and
along with it any doubt that we could smash the last great
frontier. The time barrier was within our grasp and waiting
to be broken.

DANVILLE PUBLIC LIBRARY
DANVILLE, INDIANA

BREAKING THE TIME BARRIER

THE RACE TO BUILD
THE FIRST
TIME MACHINE

Jenny Randles

PARAVIEW POCKET BOOKS

New York London Toronto Sydney

DANVILLE PUBLIC LIBRARY
DANVILLE, INDIANA

PARAVIEW
191 Seventh Avenue, New York, NY 10011

POCKET BOOKS, a division of Simon & Schuster, Inc.
1230 Avenue of the Americas, New York, NY 10020

Copyright © 2005 by Jenny Randles

All rights reserved, including the right to reproduce
this book or portions thereof in any form whatsoever.
For information address Pocket Books, 1230 Avenue
of the Americas, New York, NY 10020

ISBN: 0-7434-9259-5

First Paraview Pocket Books trade paperback edition April 2005

10 9 8 7 6 5 4 3 2

POCKET and colophon are registered trademarks of
Simon & Schuster, Inc.

Manufactured in the United States of America

Designed by Jaime Putorti

For information regarding special discounts for bulk purchases,
please contact Simon & Schuster Special Sales at 1-800-456-6798
or business@simonandschuster.com.

099889

To my mother
and the staff of the Conwy & Denbighshire
Health Authority,
who provide true support for us both.

CONTENTS

Contents

THE DAY THAT TIME STOOD STILL

The morning dawned like any other, but this was not an ordinary day. A science experiment had begun that was about to change the face of physics. It might ultimately change the entire world.

The experiment grew out of an idea first imagined seventy years earlier by two giants of science—Albert Einstein, who may go down in history as the father of time travel, and Indian scientist Satyendra Bose, after whom we name the boson, a particle that plays a key role in the construction of matter. Their brilliant suggestion was so advanced that it could not be tested at the time, as the technology needed to do so was not available. Not until the end of the second millennium would the means become available to a new generation of researchers.

The breakthrough experiment that took place that day created a Bose-Einstein condensate, a substance that is named in honor of the two men. This odd substance resembles a super atom with millions of particles smeared into one huge cloud of gas. What emerges has an undefined size and shape that's quite unlike other more mundane elements.

Einstein and Bose thought of creating such a condensate after a close study of a theory proposed by a colleague, Werner Heisenberg, a flawed genius who investigated subatomic space and came close to building an atom bomb for Adolf Hitler. Heisenberg proved that at the microscopic scale only one property could be precisely defined at any time. We might accurately measure momentum (a product of mass and velocity) or correctly gauge the position of particles—but we cannot completely measure both of these things together. The better we define one property, the less certain must be our knowledge of the other. All physical measurement is irretrievably shrouded in confusion thanks to this discovery. We have had to accept that there are limits to what we can know—not just for today or tomorrow, but probably forever.

Physicists eventually acknowledged this "uncertainty" principle of Heisenberg's, but Bose and Einstein spotted that it provided a fascinating opportunity. If you freeze atoms down to the coldest theoretical temperature—called absolute zero—all motion stops and the particles making up the atom will have no velocity. This means that momentum must also now be zero, because whatever the mass, when multiplied by zero velocity, it's bound to equal nothing. But if we know momentum exactly (even if it is zero), then, according to Heisenberg, we cannot know anything about location. In other words, when frozen to such bitter coldness a cloudy smudge of particles with no defined shape ought to form. This would be the condensate.[1]

In February 1999, Lene Vestergaard Hau, a Danish scientist working at the Rowland Institute at Harvard, successfully created a Bose-Einstein condensate and shone light rays through the resulting vapor. The misty cloud behaved precisely as Einstein and Bose had predicted seventy-three years before, and that behavior was as remarkable as the two great scientists had anticipated. Its effect on light was so curious that the script for this experiment ought to have been directed by Steven Spielberg.

As the light beam passed through the cloud, its speed started to fall from the familiar 186,000 miles per second in space to an amazing 38 miles per hour as it crawled through the mist. Rays of light normally move so fast that our senses cannot see their motion, though they do get impeded slightly as they pass through solids or gases. This itself is not odd. What was extraordinary was that these light rays were no longer fast enough to circle the Earth in less than a second. They were traveling very slowly, and it was possible to imagine catching up with them. But what would happen to something that catches light? This question, which was once in the realms of science fiction, was all of a sudden quite real.

As Hau's team worked to improve their method, Ronald Walsworth and Mikhail Lukin of the Harvard-Smithsonian Center for Astrophysics made successful modifications to the experiment. In January 2001, after creating another cloud of condensate, they shone *two* beams of light through it. These interfered with the particle waves inside the smudge, just as two stones dropped into a pond will form separate ripples. The ripples then spread out across the water and when they meet, change the wave patterns of one another.

Inside the mist of condensate the transmitted light beam moved ever more slowly—just as Hau had shown that it would. But the second beam—a laser—was the key to what happened next. The team ensured that its frequency was slightly different from that of the first light beam, allowing manipulation of the interference pattern from the interacting ripples. This caused the light rays to slow beyond the levels that Hau achieved. The results were unprecedented.

Slowing light from its normal very fast speed might seem to be of mere technical interest. But it's actually quite significant. Light conveys the flow of information that defines all human experience. What we see, know, or understand about the world comes from the data brought into our senses on light rays and similar waves of energy. Light, quite literally, creates time as we experience it from moment to moment. Find a way to alter light's motion and you begin to defeat the time barrier.

In the Harvard-Smithsonian experiment one of the two beams shining through the condensate acted as a flashlight to illuminate the mist—making it transparent. The second light beam could then be observed wending its way through the freezing vapor—describing a path just like any ordinary ray of light might do when entering a room through a gap in the curtains. But this light ray was moving incredibly slowly.

Light speed, once thought to be infinite and untouchable by scientists, was now being tamed by human intervention. And in that moment a new realm of possibilities opened up—including the keys to a time machine. Science thrives on the challenge of any environment, conquering air and space in this way, but until now the persistence of light speed seemed to deny any realistic hope of traveling through time. Since we all experience essentially the same progression from past into future as a steady rate, there was no obvious way to try to power through the temporal medium either faster or more slowly. Yet the Harvard-Smithsonian experiment proved that light speed could be manipulated and brought down to a level where you might beat it in a race. And anything that runs faster than light takes the first step towards traveling through time.

When Walsworth and Lukin analyzed their results, they saw that the speed of the light passing through the condensate had surpassed the record set by Hau, which she had wryly noted could be outstripped by a good rider on a bicycle. In their experiment, the speed of light through the cloud had become virtually imperceptible, and a moment approached with awesome implications. The beam of light moving through this misty cloud reached zero miles per hour. Its information flow had ceased completely. Time had been stopped.

If you stop the flow of light conveying information, then its data transfer will stand still—because no news regarding past or future will be carried any further along its course. Stop time and it is no longer impossible to imagine traveling through the frozen temporal landscape. Will past and future spread all around us like a foreign country waiting to be explored? Could we dream of finding the right vehicle and a road map that would plot a course into its midst— perhaps a course taking us into the past or the future?

A ray of light had been stopped dead in a laboratory and along with it any doubt that we could smash the last great frontier. The time barrier was within our grasp and waiting to be broken.[2]

Introduction

THE RACE

In September 2003, General Wesley Clark made an unsuccessful bid for the Democratic candidature of the United States presidency and in the process addressed a group of New Hampshire voters with a very strange rallying cry. Clark suggested that humanity should aspire to travel faster than light and that this goal could form the catalyst for a dramatic new race between competing cultures. The ultimate success of this mission would allow voyages between the stars. He may not have realized that it would also do something else much more dramatic.

Some media sources smartly interpreted these words as a suggestion that the United States government should fund a search for time travel—although Clark did not specifically argue for that. Weird as it seems, such a conclusion was not illogical because—from our understanding of physics—if you travel faster than light, then you can overtake the flow of events that light happens to transmit. Since the passage of these events forms what we interpret as time, then by traveling faster than light you ought to travel through time. Spaceships that outstrip light speed are always going to moonlight as time machines.

In suggesting a race of this sort Clark was echoing the wonderful speeches made by John F. Kennedy in the early 1960s concerning space travel. Clark saw his twenty-first-century quest as a uniting force, not a political battle. Even with the inevitable element of competition it would give mankind a purpose that could benefit all, with countless possible spin-off advantages. While it may, or may not, be technically possible to beat the huge velocity of light in space, we should make the effort, the general concluded.[3] Ironically, he likely did not know that we had already succeeded in making light stop dead inside a misty condensate.

Clark also probably did not know that a race to build a time machine has been going on since at least the Second World War. This new endeavor differs from the space-race, because there is not the same degree of national competition, although science can thrive on rivalry between researchers who aim for the same goal. The primary battle in the "time race" is between conflicting scientific theories and pits those who would challenge the frontiers of nature against those terrified that by doing so we might upset the balance of the cosmos. This race may be the greatest scientific enterprise in history and offers the prospect of great riches to whoever wins and, no doubt, will bring massive benefit to the nation that sponsors this research. It will touch the lives of everyone on Earth.

The time race has involved scientists from all over the planet, as well as the occasional maverick inventor keen to scoop the prize. Each one has attempted to produce a device that will carry a human being into the past or the future, or possibly do both. Some feel we may have to settle for less direct methods of time travel that do not involve human chrononauts (a word derived from Chronos—the ancient god of time—and used for would-be time travelers). Others do not see the need to impose any such restrictions. Either way, the battle lines have been drawn and the competition has gained enormous momentum in the past few years, to the point that many former skeptics now admit that the time barrier is ready to be shattered.

This book will tell the story of how science has wrestled with understanding the nature of time and how this has inspired the hunt for a time travel device. The story is sometimes strange, sometimes surprising. But all of it is true.

The concept of a time machine has now thoroughly infused popular culture. Movies, science articles, and novels—such as Audrey Niffenegger's romantic tale *The Time Traveler's*

Wife—that treat the subject seriously are everywhere. In Britain, BBC (British Broadcasting Corporation) television brought back its long-running time traveling hero *Doctor Who* for a multimillion-dollar series because of growing demand. The theme has captivated the public imagination like never before. Many sense that the day of the time traveler has arrived.

In telling this history of the time machine, I will explain why this fascination has intensified and will reveal the current state of this ongoing race. How close are we to achieving this extraordinary possibility? What will we be able to do with time travel if and when we succeed in making it happen? How seriously should we take the claims of those who say that they have already broken the time barrier? Will we ever be able to have a time machine for personal use alongside the family car?

The race to conquer time has spanned several hundred years, but it is only since Einstein that there has been any legitimacy to the science behind the efforts to time travel. Until then, and in many ways long afterwards, it was widely considered the province of crackpots to talk about constructing a time machine. These days it is the subject of erudite discussions in physics journals and research grants allowing genuine experiments to be conducted at prestigious institutes. There is excitement in the air around many a university campus, and that causes scientists to believe that this amazing concept is now within our grasp.

This optimism is not universal, however. There are plenty of scientists who abhor the whole issue of time travel. To them it is their worst nightmare—and they are not just being melodramatic. Time travel brings massive problems that threaten to blow huge holes in cherished theories and long-standing concepts about how the world works. The traditionalists are not letting go without a fight. So, those in the race to build a time machine face an insidious enemy from within, adding to the inevitable guffaws from less understanding outsiders who see their work as the products of cranks. Some of their colleagues are determined to prove that time travel cannot be possible and will stop at nothing to outlaw what to them are absurd ideas and nonsensical experiments.

Physicists are not horrified by time travel just because it is the product of a lively imagination. It may fill endless episodes of assorted adventure TV series, but time travel has a solid foundation in what we know about the physics of the universe—and that is the whole problem. It is more and more apparent that the laws of physics are designed to facilitate time travel, not to deny it, and many scientists are worried about that fact because it challenges the very fabric of the universe.

If the concept of time travel were merely speculation then it would be of little importance to mainstream research. Science is not easily persuaded by extreme ideas, and scientists know the distinction between the romantic dream and the practical reality that offers a prospect of technological pay dirt. However, time travel is not just a dream and the pay dirt on offer seems to be quite real.

For the past century the idea has been gradually transformed from the wide-eyed visions of Victorian novelists into serious laboratory experiments that are seeking to make it happen. The charming fiction of H. G. Wells, who first wrote about a time machine piloted by a daring hero back in 1895, has morphed into fevered debates between world-renowned researchers arguing over how it can be made to work.

The United States landed on the moon after a space race stimulated by the tensions of the Cold War. It benefited the economies of both the United States and the USSR by creating tens of thousands of jobs and inspiring many products that we use in our daily lives, such as medical techniques that could only be developed in the zero gravity of space. A Nobel Prize may be more than enough to persuade many of the would-be chrononauts to keep on trying to win this new race, as most of the cultural battle lines have disappeared with the breakup of the USSR. Even so, national pride is likely to play some part in fueling an event of this magnitude. Indeed, as this story unfolds, the breakthrough experiments, theories, and bold plans to travel through time have spanned the globe, putting one nation ahead and then another in this fascinating contest. While the United States has sponsored many of the most dramatic developments in the past few years, the first claims of major success are coming from an unexpected source—an old enemy, in fact—perhaps eager to secure this chance for a place in history. Meanwhile, there are independents constructing their own prototype time machines who have their own, sometimes deeply personal, agendas.

Indeed, the history of human ingenuity, from the discovery of fire to today's remarkable feats of genetic engineering, is just a series of steps along that same path to enlightenment. We are strengthened by our desire to do the impossible, not defeated by the thought of its many difficulties. But time travel is of a wholly different order to the journey that led to the modern computer by way of the abacus and calculator. It is important to try to understand why this difference in scale exists and why it is so shocking to science.

The problem is that time travel should not be able to happen, according to the classical way of viewing the universe. By all logic, as the nature of reality has been understood across centuries of enlightenment, traveling through time should be absurd. Yet modern physics, in ways that experiments are proving day after day, shouts loud and clear that time travel is not only possible, it happens all around us as part of nature.

It is no surprise that many physicists are having bad dreams about hordes of time travelers traversing the byways of the universe, nor to learn that scientists have split into warring factions regarding the awesome implications of this field.

On one side we see those who suggest that, since modern physics says that time travel ought to be possible, then there must be something wrong with modern physics. Or, more correctly, that there must be something incomplete about our understanding of this subject that makes time travel seem to happen when it surely cannot. The renowned Cambridge don, Professor Stephen Hawking, was one of the first to take this stance (although—like many—he has moderated his views in the past few years). He began by arguing that there must be an as yet undiscovered rule within the workings of the cosmos that will deny time travel—a restricting force to prevent this travesty from wrecking the status quo. Nobody knows what this rule might be, and in truth his suggestion is little more than a cry for help! Nonetheless there is a search for his theoretical "impossibility" mechanism that is galvanized by the desire of the skeptics to find it, although at the same time it's complicated by the fact that nobody knows if it is really out there to be found.[4]

There is also now a growing band of physicists who have adopted the bolder perspective. They argue that, since our modern views about science seem to show that time travel can occur, then let us stop worrying about why we find this outcome horrifying and get on with trying to develop a practical method of traveling through time. They are by no means the first to make attempts to build a time machine. For over a century eccentric inventors have been trying. But the new band of physicists is the first with a real chance of success.

So why is time travel still regarded as being absurd by so many more cautious commentators? It has all to do with the question of the temporal paradox, and the best way to illustrate such mind-bending riddles is with an example. Let us consider two hypothetical brothers, both scientists who are determined to perfect a time machine. We will call them Professor Fred Cleverman and his brother Ed. Their exploits will dramatically illuminate the problems.

Being a romantic soul Fred, as the first to succeed, decides to use his machine to travel back fifty years to secretly observe his own parents on the night they fell in love. He hovers out of sight in his floating time machine (time machines would have to fly to avoid materializing inside an object in their path in the past). As he observes the youthful couple who will become his parents, they seem to be getting rather amorous. So Fred starts to blush and tries to press the reverse lever to back away before one thing leads to another. Unfortunately, as this is a new machine and he is not familiar with the controls, he presses the accelerator in error and promptly speeds his machine forward onto the roof of his parents' Chevy.

Fred survives the impact because twenty-first-century materials are tough. But the fate of the 1950s motor vehicle is not so happy. It is crushed, tragically along with its occupants. Professor Fred looks down in horror at what he has done and prepares to fly away, back to the future, knowing that in his enthusiasm he has just killed his own mother and father before they can start to conceive him.

There is no problem here, you might think. All Fred needs to do is go back in time to a period just a few minutes earlier, hover alongside his own machine as soon as it arrives in the past and warn it off before the accelerator is pressed. But sadly, when you think this through, it quickly becomes evident that this cannot work and we discover why we call such an event a paradox.

Professor Cleverman cannot go back to warn himself for one very simple reason. The moment that he squashes his parents, they cannot survive to *be* his parents. As such he is never born, does not invent a time machine, and cannot have flown back and accidentally killed them in the first place.

So is Fred born or not born? Does he travel back in time or not? Does he kill, or not kill, his parents? These are the migraine-inducing questions that you face when confronted by a temporal paradox. There is no answer to these questions because the whole experience is a minefield of logical impossibilities. If they die, he is never born, so does not go back and they do not die. If they live, he is born, does go back, and they die as a result. It's no wonder physicists are having nightmares!

Introduction

Yet, however confusing are the adventures of the hapless Fred Cleverman, the story of his brother, Ed, is even more disturbing. As a boy, he hit upon his idea while reading through obscure texts about temporal physics in a local science journal. Nothing more was heard about the theories or their author, but they inspired young Ed to build a time machine following decades of effort.

Unsurprisingly, his first time trip was to meet the unknown scientist who wrote the thesis that proved so inspirational in order to reveal that his theories would one day be vindicated. Unfortunately, there turns out to be a problem. Ed gets back to the 1960s, very discreetly, of course, since openly proclaiming yourself a time traveler is like asking for a ticket to the funny farm. Once there he discovers that the article is not in the files of the science publication although Ed knows it should be in their very next issue. What is worse, nobody has heard of the man who has written it.

Ed now faces a huge dilemma because, if the article never appears, he can never have read it and thus gone on, thanks to its ideas, to build a real time machine. He, therefore, could not be back in the past searching for the article or its author as he was now doing. Afraid that he might be about to disappear into a vortex of nothingness—he saw it happen once on an episode of *Star Trek*—Ed hits upon a bold plan. He writes the article himself, since he has memorized it word for word, and submits it to the publication using the name of the apparently nonexistent scientist whom he had always held up as his hero and mentor. The article appears and Ed goes home to the future sure that all is well with the universe.

Except that all is far from well, because what has just happened is a serious breach of the rules of nature. By attempting to cleverly defeat the paradox he faced, Professor Ed Cleverman has shattered what physics calls the law of "cause and effect"—a near sacred tenet that argues quite simply that an event is *always* preceded by the cause that makes it happen. A moment's thought will show why.

You pull the trigger and then fire a gun. A bullet does not pop out of the barrel followed by your decision to pull the trigger because you have seen it emerge. You strike a bell and the bell rings. You do not hear a bell sound and then realize that you had better strike it in order to make that event actually happen. What if the person next to the bell chooses not to strike it? By all common sense the bell, of course, would not ring. But it *has* just rung. So it would appear that again there is a paradox, meaning that you *must* strike the bell in order to ensure that the universe remains in order. But then every tiny event would be predestined and there could be no free will, when experience tells us that there self-evidently is.

This is a very simplistic demonstration of cause and effect. In reality there are detailed mathematical rules that cover it, but they all say much the same thing: You cannot have an outcome that precedes the cause that made it happen.[5]

Ed Cleverman may believe that he has put things right by ensuring the existence of the building of his time machine. Indeed, he has rather ingeniously looped back on himself to make the effect become its own cause. But in doing so Ed has solved one paradox by creating another—the production of an impossible series of events. It seems ludicrous to be able to reinvent the past by using knowledge from the future. Otherwise where does such a paradox end? This idea has been the basis for countless inventive novels and on its own seems enough to make time travel nonsensical.

When faced with the twin tales of horror that Ed and Fred Cleverman reveal to the physicist, there is only one seemingly sensible route to follow. That route is precisely where most scientists have fled—in arguing that in order to prevent these insane outcomes from twisting the fabric of the universe then time travel must be impossible, whatever physics currently says. If you have millions of time travelers creating paradoxes like these then the nature of reality would be in chaos. Time meddlers would be everywhere.

Unfortunately, modern physics presents us with its own version of the paradox by showing that, however odd these bizarre repercussions seem, the mechanism to create such chaos does genuinely exist within the laws of nature. Our modern knowledge of how the universe works not only supports time travel, it illuminates the path towards allowing both of these outrageous consequences stemming from it.

Of course, scientists, when faced with this sort of dilemma, have two options—just as any human being does. They can face the issue head-on (seeking to discover how time travel might work and, hopefully, eliminate the paradoxes while deciphering its nature) or they can run and hide. In a sense those who say that science must be wrong if it allows for time machines adopt the more defensive standpoint.

This book will follow the ideas and experiments of science throughout the past one hundred years that have led us towards a conclusion of this ongoing debate. The twists and turns will seem like a plot from a Hollywood movie, as grand theories emerge, are vindicated, are shot down in flames, and other ideas emerge from their ashes. Yet, even if the weird consequences that emerge look like fiction, the race to build a time machine is straight out of science fact.

That's not to say that fiction—particularly science fiction—doesn't have a role. It very much does. Many modern writers of time travel stories are working scientists who use drama as a way to test bed their ideas. There is a subtle interaction between the factual experiments and the fictional tales that ponder the results. This has gradually brought the race ever nearer to its stunning conclusion. For we are close—very close—to commissioning that first time machine.

Pre-1895
THE DAWN OF TIME

To run any race you must know the course. To build a time machine you need to know what time is, just as you cannot fly without knowing the nature of air and aerodynamics. But understanding time is easier said than done.

A celebrated Zen riddle asks, when a tree falls in the forest and nobody is around, does it make a sound? This riddle can probably be applied to time. Would there be such a thing as minutes or years if no human beings could experience their passage?

This seems to be a very odd suggestion, but the nature of time is very strange. Indeed, it is a real puzzle for science. It forms an inescapable part of our lives yet cannot easily be defined. It has fascinated mankind since we first learned to communicate, but there have been no clear answers about its nature. Indeed, some great minds have argued that its measurement is purely a human invention.

Greek philosopher Zeno showed the problem when he tried to define a small unit of distance. To catch a tardy tortoise you can easily run twice as fast and halve the distance between you and the animal in a set period of time. But if you keep on halving the distance that gap will never equal zero, because half of something is always going to be a finite number, however small. But if there is always a gap between you and the tortoise it is impossible to ever catch up with it—a conclusion that we know to be absurd by practical experience, even if we have never actually chased a tortoise. A faster runner will always catch a slower one, sooner or later.

Time is intimately involved in this discussion—since speed is a measure of distance traveled in a set time. So we can apply Zeno's thinking and divide a second into smaller and smaller pieces. If we keep breaking down this gap, making the units half as long as the previous one, then there will always be a finite length for any moment that we can measure. But if that moment has any size at all, then part of it must be in what we think of as the past and part of it in the future because it will take time to pass any mark or point. We call this tiniest measurable moment "now" and say that it separates past from future. Yet how can it separate anything if parts of it lie simultaneously in both past and future?

Arguments still rage over the meaning of this curious riddle. Is it a fallacious argument—like the one concerning the tortoise? After all, it may look impossible to catch up with the animal but clearly we know that it is not, so the riddle is flawed in its execution. Others suggest that there may be something even more profound in this realization about time first made 2,500 years ago. Is the reason that we cannot clearly identify a moment that is neither past nor future a hint that past and future are a product of human imagination? Is the universe fundamentally timeless and is the distinction between past and future just an illusion brought about by our limited capacity to visualize the cosmos?

Virtually every human society that developed a culture has speculated in similarly bemused ways about the nature of time. The Greeks defined it as a measurement of intervals, which could be of long or short duration. As far back as 350 BC Aristotle had realized the implications of the Zeno paradox. But he had no better answer, and this choice to divide time into basic units, mirroring many mundane things that form a sequence, such as the human heartbeat, allowed for the creation of sundials, water clocks, and eventually mechanical clocks. We gained a feeling of mastery over time by recording it with increasing skill and so it came to be a powerful element in our lives.

St. Augustine, many centuries later, was a little bolder and dared to ask the question—what was God doing before He created the universe? If time was born along with the matter in the universe, as the Bible suggests, then was there any time before that instant, or is God somehow also to be considered timeless? Intriguingly, this question largely foresees modern scientific concerns about how the cosmos was first created—the subject of intense debate between physicists and astronomers.

There are two basic theories. One is the so-called Steady State idea that the universe has always existed in its present form, perhaps even made by God. British astrophysicist Fred Hoyle championed this theory though he also invented the name given to the rival theory—the Big Bang. He hoped that such a daft name would ridicule this alternative concept that says that everything in the universe emerged long ago from one single, tiny point that exploded outward and has gone on expanding across billions of years. But the Big Bang theory has gathered strong evidential support from modern science and is the widely accepted view today. Hoyle was proved wrong, but ironically, his name is attached to the theory that he so detested—the ultimate insult. Physics has had to conclude that time somehow began when the universe started to expand and before that instant there was neither matter nor time.

However, it was by no means clear to Renaissance thinkers that time emerged from the birth of the universe.

Nor did they even accept that it was an essential require-
ment to make the laws of nature work. Indeed, the more
that science began to comprehend these rules, the more it
became aware that time, in our experience moving from the
past into the future on a perpetual one-way journey, is not
a prerequisite. In fact, virtually every law of physics seemed
to work just as well if time flows backwards, moving from
the future into the past. This realization enhanced suspi-
cions that time might be a convenience of mind that made
us see things as we do rather than a necessity of nature.[5]

Different societies have other concepts of time and it is a
mistake to imagine that our modern Western perspective,
dominated by timetables and cell phones, is the only way
to see things. We have grown up with this one version of reality but there are other, equally valid interpretations. The
dreamtime, for instance, is an aboriginal concept still widespread in native Australian culture. It could not be further
removed from twenty-first-century thinking and is extremely difficult to even translate. But, in essence, it regards
past, present, and future as coexisting in a timeless void or
hidden dimension beyond the range of our normal perception. For that reason, in dreams and other states of consciousness where we lose touch with the normal sense of
awareness, we enter what is in effect another reality where
things that once were, still are, and where things that will
someday be, have already become.[6]

Time spans the infinity of the cosmos and the tiniest

 DANVILLE PUBLIC LIBRARY
DANVILLE, INDIANA

moment that we can record. But it may not even exist. No wonder it is such a riddle. But it is important to follow the manner with which science has attempted to come to terms with time, piecing together its nature through a series of grand theories and experiments. For these are the stepping-stones upon which today's plans to build a time machine are all based.

Throughout the Renaissance, as scientists began to understand the nature of the physical world, there was an uneasy truce between what mattered to most human beings and the things that interested physicists. Galileo and Newton showed that all the planets of the solar system, including the Earth, rotate around the sun in a wonderful cosmic ballet. Their paths could be mathematically defined, to the point that Newton even argued that God created the universe as a vast clockwork machine that allowed everything that would ever happen to be mapped out into perpetuity. God had wound up the machinations of the cosmos and let it loose for mankind to discover its properties. By doing so we could make stunning calculations far into the future, because the speeds and times of the orbits of these planets could all be precisely delineated effectively forever.

It was these calculations that allowed NASA to work out how to send Apollo spacecraft to the moon, using sums that Newton could have easily done for them. The same rules allowed the rescue of fated mission, Apollo 13, sending it like a slingshot around the lunar surface and heading back to Earth thanks to the mathematics of the universe and its timeless precision.

However, as these findings seem to prove that ticking clocks were defined by the distant motion of bodies in space, science also found itself in open warfare. It battled religion, fearing that the mathematics of nature might replace the edicts of God. And it battled ordinary people who had always gauged time in simple ways—from observing the seasons, the growth of crops, and the calendars decreed by the church. Now scientists were saying that the only true way to measure time was to accurately describe how the Earth revolved around the sun and the exact time it took for our planet to rotate on its own axis. We had only ever been able to make guesses about such matters before and had inevitably miscalculated to some degree. Scientists wanted to put right those centuries-old mistakes and rearrange the timetable of our lives so that it was in balance with the motions of the universe.

In the 200 years leading up to the nineteenth century, ordinary folk were asked to rethink how they should now judge time. For centuries the year had been calculated as having 365 days plus one quarter of a day, hence the extra "leap year" day every four years, but this estimate based on the Earth's orbit was only approximate. As time had passed the year had slipped out of phase with the way our planet truly moves around the sun, and did so a little bit more each year. So by papal edict in 1582 the error was corrected and 11 days were dropped from the calendar. Such was the opposition to meddling with time that this "Gregorian" Calendar found favor only after a long period and with some decidedly odd consequences.

For instance, the area surrounding the city of Strasbourg accepted the decree immediately and changed over in November 1583. But the city itself stuck to the old calendar for another ninety-nine years—meaning that when it was New Year's Day in Strasbourg it was already the middle of January just a few miles away. The chaos that resulted is obvious, not to mention the apparent time traveling—by crossing the city line, you could walk "into the past."[7]

In Britain workers protested that eleven days would be stolen from their earnings if they agreed to the plan imposed by Rome. Such "time riots," as this clash between science and the masses was dubbed, shows just how much concern was being expressed by the ordinary, then generally uneducated, person about any attempt to play with our long accepted way of viewing time. They delayed the introduction of the correctly aligned calendar in the United Kingdom until 1752, almost two centuries after much of Europe.

The old ways of thinking about time have not entirely gone away. For instance, on the Isle of Man in the Irish Sea (the world's oldest continuously operating parliamentary democracy) a ceremonial reading of laws to the public is still held at Tynwald Hill each year. It occurs on what would have been midsummer had those 11 days not been expunged two and a half centuries ago.[8]

A crucial moment in the understanding of time came with the ability to measure the speed of light—although, when this happened it was not apparent that there was even a speed to be measured.

It had long been assumed that all objects emit "rays"— which Newton suggested to be streams of particles—and that these traveled in straight lines to reach the eyes. Our eyes absorbed the rays and became "excited," thus seeing the object. Because the process happened so swiftly it appeared to be instantaneous. We could detect no varying time lag between viewing our hand held in front of our face or the moon, which is very far out in space. So it was reasonable to assume that light traveled instantly.

Research by Isaac Newton in the late 1600s, using prisms to split light and unravel its makeup, led to the underlying truth. Light does indeed convey information to our eyes. It acts as the yardstick of all events, perhaps even the creator of our perception of time. How fast it moves is crucial, because this determines whether the past really is gone forever, as is widely assumed. If light flowed like a river, which was then the prevailing belief, then once you were swept past any point in your journey on the way upstream all you could do was keep on moving forward. But if light has a speed, like a current on a river, then perhaps you can find a way to travel downstream at a faster rate than the current and thereby return to a place that you have sailed past before. That other place would be the past.

Since general experience dictates that we do not have the ability to revisit the past, except in our memories, this furthered the belief that light came to our eyes instantly, meaning there was no speed that we could ever hope to overtake. But that opinion proved to be wrong.

In 1676 the Danish astronomer Olaus Roemer (1644–1710) devised a clever experiment. He used the laws of planetary motion that Newton had set down, the telescope that Galileo had developed for astronomical observation, and the moons of the giant gas planet Jupiter that Galileo had discovered orbiting majestically around their parent. Bringing all these discoveries together allowed Roemer's test to expose the rules of time as being quite different from those commonly held.[9]

Every now and then these moons pass behind the huge mass of Jupiter because its body blocks them from our view on Earth as they orbit. If light speed was infinite, then the time taken for this period of eclipse should always be the same. But Roemer found that it was not the same. In fact, once he had tabulated enough measurements he could easily discern a pattern behind them.

The eclipses took longer to happen whenever the Earth was moving away from Jupiter because of the relative motions of the two planets around the sun. At different times these same relative motions caused the Earth to move towards Jupiter and when it did so, Jupiter's moons passed behind the planet a little bit faster. Very cleverly he then deduced that the difference in time for this same event was the result of light having a finite speed. And that speed could now be worked out from these observations.

When the Earth was moving away from Jupiter, the beams of light conveying the image of this eclipse had to travel farther to catch up with our planet as it sped away. So the light took longer to reach us and the event seemed to last longer. The reverse happened when the Earth was moving towards Jupiter. Light rushing to bring the image to our eyes got here faster because we were moving towards it, closing the gap. Consequently the eclipse seemed to have a shorter duration on these occasions.

Roemer had only roughly accurate calculations about planetary sizes, orbital speeds, and distances to work with to let him figure out the speed of light. He argued that it was about 140,000 miles per second—so fast that it was understandable that the human eye had always assumed it to be instantaneous. The real speed of light, with accurate modern measurements, is closer to 186,000 miles per second. Light speed is almost a million times faster than sound, hence the obvious discrepancy between hearing the sound (thunder) and seeing the light (lightning) caused by the same physical process during a storm.

That light possesses a measurable speed was unexpected, yet it would be critical to how we interpret time. Roemer's calculations came in an age when humans were constrained by the thought that the speed that a horse can gallop was about as fast as you can go. However, now that Roemer had shown the true picture of light's extraordinary speed, the frailty of human capability was exposed. We were not masters of the cosmos and there were things that might prove to be beyond our comprehension. We could start to question what we had considered beyond doubt and imagine building something that might fly as fast as light. What would happen if we shattered this light barrier? For the first time in history, that became a legitimate question.[10]

Philosophers started to wonder about these matters. To thoughts about such major quandaries as the existence of God or free will were now added bold thinking about manipulating time. This was the birth of science fiction, and the time travel story is one of its staple features. But this would not have been possible without Roemer's experiment, because it was his discovery that allowed for time travel to even be considered as a theoretical possibility.

Indeed, these philosophers recognized something that science had missed. There were two types of time. One of these was the real time that existed as a fundamental property of the universe (according to Newton) and was controlled by the speed at which light conveyed signals (as measured by Roemer). Philosophy added to this the subjective experience of time. Time flies when you are having fun, the adage says—emphasizing this human subjectivity.

Summers of childhood seem to last forever, whereas adult years rush past much faster than you would like. There were many other examples of this intangible nature of time, allowing strange questions—such as whether we could change our interaction with time just by thinking about it differently.

This has inspired many intriguing science fiction stories, such as Richard Matheson's haunting old tale *Bid Time Return*, which was filmed in 1980 as *Somewhere in Time*. Here the hero falls in love with a woman from the past and mentally time travels just by removing every attribute of the modern world from his environment and sensory perception. By soaking himself in the sights, sounds, smells, clothes, furniture, and trappings of another age he gets handed the road map to take him there. Sadly, he is snatched back to the future when he discovers a modern coin accidentally left in his pants.

Science, of course, scoffed at such romantic notions. Time, to scientists, is not open to human intervention in this simplistic manner but is a real property of the cosmos. Yet whether we can or cannot think ourselves into the past, we all do travel into the future with each passing moment of our lives. Washington Irving in 1850 cleverly realized that when we fall asleep we lose touch with any sense of time and awake perhaps a day into the future without any apparent recall of the journey. Like all good science fiction, his novel *Rip van Winkle* just took that insight and exaggerated it into a situation that we can imagine might really happen one day. What if you fell asleep not for a few hours or a day but for years? Subjectively, you would seem to have awoken into a future age, as if you had time traveled there. But, of course, this was time travel gained the hard way—at the expense of the limited number of days left in your life.[11]

Other Victorian novelists attempted different ways of taking people into the future, by being accidentally frozen in ice or trapped in a cave for thousands of years. If the normal deterioration of the body slowed down in the process, then longevity would bring about the experience of time travel. But these were not real speculations on time travel, nor even on time, but rather the use of time as an artistic medium to paint a picture of human actions or social circumstances.

The inventive novelist Samuel Clemens, better known as Mark Twain, created such a social satire with his 1889 story *A Connecticut Yankee in King Arthur's Court*. This was the first time travel story to involve travel into the past. Twain's hero travels back from nineteenth-century America to pre-medieval Britain but as an expedient to the story rather than as scientific speculation. Indeed, there really is no science in this fiction. The purpose is to contrast society across the ages, and to rapidly expedite the time jump an electrical storm catapults the traveler. There is no attempt to explain how this might be possible, but by the 1880s physics was making exciting discoveries about energy emissions such as lightning. Twain no doubt saw this as an appropriate basis for making the impossible come true.[12] However, this was also fortuitous because electrical power has since become a useful ally in the scientific battle to surmount the time barrier.[13]

Six years after Twain's story, the young English newspaper columnist H. G. (Herbert George) Wells wrote the first novel about an actual machine that could travel at will through time. Despite his primary occupation, Wells had a science background, being a graduate of Imperial College, London. This would stand him in very good stead.

The scientific knowledge of his day was the basis for many of his stories, which predicted, for instance, aerial bombing and nuclear warfare, but *The Time Machine* was to prove so influential that its centenary was even celebrated with a special stamp in the United Kingdom. Wells's romance describes how an inventor creates a device to travel through the "fourth dimension." He was aware that physics was starting to regard time as having a dimensional nature—even though the full impact of this concept was yet to be realized. Like all good writers, he kept up with the latest research and asked: What if we could move through the time dimension just as easily as we travel through the spatial realms?

Such an idea was perhaps prompted by powered flight that in Wells's time was about to conquer the skies and at long last tame the up-down spatial dimension. So if time was an added dimension, why not build a machine to fly through that region as well? The extraordinary possibilities opened up by his novelistic invention of a time machine meant that Wells had created not merely a story, but a new human aspiration that would grow in the minds of everyone who read his novel. That included real inventors and young scientists who would later seek to make his idea a reality.[14]

In the enthralling 1979 spin on his story, filmed as *Time After Time*, the real life Wells is argued as having actually invented a time machine, which is promptly stolen from a dinner party that he was hosting back in 1893. Unfortunately the thief is a doctor suspected by many to have been the real serial killer "Jack the Ripper"! Jack flees in the machine to San Francisco in 1979 and recommences his old habits, pursued by Wells after the time machine emits its "homing" signal and returns to his lab. Taking it to 1979, Wells has a series of well-observed adventures, defeats Jack, thereby explaining why the killer was never identified in his own era, finds a modern American girl, falls in love with her, and both return to 1893. She becomes the woman who would be Wells's true life wife; he then destroys the time machine and writes all of his research up as a novel and the past becomes exactly what we know it as today—without violating too many laws of physics.

Ironically, this idea is not entirely fanciful—in the sense that the real Wells did indeed try to build one of the first actual time machines. It was not as grand as the one in his novel or the subsequent modern movie version. Indeed, it is probably best described as a time travel *simulator,* like the one found these days at some theme parks where you can go on a thrilling ride through time courtesy of state-of-the-art technology. Motion simulators allow your body to react as it would in flight through the air as you are immersed within the imagery of a journey. This tricks the body into believing that the experience is real.

This is just what Wells attempted to do in 1895 along with pioneer movie producer Robert Paul. They wanted to build a movie theater that would give the audience the illusion of traveling through time by putting them inside the action that was projected on screens all around them. They lacked the machinery that would have made it very convincing—indeed, cinema itself was still at a very primitive stage. But it was a far-sighted plan that, unfortunately, never got off the ground.

Nevertheless, Wells's influential story anticipated the direction of our knowledge about time and space. But did his prophetic expertise extend to the prospect of a real time machine?

There is always a fine line between writing science fiction that becomes insightful and stories that just look silly after a few decades of progress. One need only view some of the far future technology dreamed up in the original episodes of the 1960s TV series *Star Trek* to see this in action. What looked like gadgetry from the twenty-third century to people forty years ago, already seems old hat to us in the early days of the twenty-first century. The number counters on board the starship Enterprise feature no LED displays, as this technology was not predicted during the 1960s. Instead, lumbering manual counters were imagined to remain in use and yet these are already more likely to be found in a museum! Futuristic computers were also imagined to be clumsier by many orders of magnitude than those that are in regular use by millions of children at home today.

It is simply not possible to predict the future of technology beyond the next few years because science is always going to provide surprises. We have many things in daily use that were not foreseeable—such as lasers, microchips, and cell phones. And lots of things we do not yet have, such as flying cars, that most futurists in the recent past expected to exist in the world of today.

Time travel could have proved another blind alley. Fortunately for Wells, almost as soon as his vision was in print, physics discovered that time travel was a genuine possibility. We have not stopped trying to make Wells's concept of time travel into a practical reality, and just a century or so later we are actually on the verge of making his dream come true.

1895
THE FIRST
TIME TRAVELER

H. G. Wells might have been the first person to imagine and then try to create a time machine, if only a simulation of one. But ironically, 1895—the year when his great novel appeared and he drew up plans to construct his time-traveling movie theater—saw the first real journey through time. Or so it did according to the originator of a decidedly odd physics experiment.

The claimant was a renowned scientist of the age, who has given his name to the unit of magnetic flux, but his story of decades of research reads more like science fiction.

This peculiar genius was Nikola Tesla, whose bizarre experiments ranged from insightful inventions still used today to some of the most fanciful allegations and terrifying attempts to use manufactured bolts of electricity. Tesla's work has been consulted by intelligence agencies and even reputedly forms a hot item on the reading list of certain terrorist groups in today's frightened world. He has become the unofficial hero of conspiracy theorists who argue that Tesla's biggest successes were covered up to prevent governments from going bankrupt. The truth has probably more to do with the way that he courted attention (today he would be a darling of the chat show circuit and would probably lecture at oddball UFO conventions). His genuine scientific pedigree did little to endear him to his peers, who thought the man at best eccentric and at worst unhinged.

Born in Smiljan, a war-torn part of Croatia, Tesla attended the University of Prague but soon moved to the United States. At the age of just twenty-eight he began to work with Thomas Edison, who was at the peak of his success as an inventor at the time, developing such useful items as the lightbulb and the phonograph. During this period of cooperation Tesla also came up with his own most enduring inventions—including fluorescent lighting and the AC generator, most of his achievements being snapped up by George Westinghouse for use in his power supply empire. Some of what Tesla discovered in those years remains the basis of domestic electricity supply over a century later.

But the honeymoon was a short one. A huge dispute arose between Tesla and Edison concerning the best method of bringing electric power into the home. Edison favored the use of DC (Direct Current); Tesla was adamant that AC (Alternating Current) was superior. Tesla reasoned that if you touched a live wire carrying DC current, it caused the muscles to contract, thus making your grip tighter and forcing continued contact. This, to him, made DC much more deadly than AC.

Edison strenuously disagreed and chose to conduct a gruesome experiment to test Tesla's convictions. He paid some street urchins to collect stray dogs, placed one of them onto a metal plate, and promptly electrocuted the animal—noting that it was frazzled but not tossed clear even when using AC current. With this he had hoped to show that his ex-colleague, now bitter rival, was wrong to believe that AC current was any safer than DC current.

To preempt arguments that dogs were not the same as humans, Edison even attempted a few experiments with larger mammals—including the public electrocution of an elephant!—but thankfully he stopped short of the ultimate test. Some authorities were less squeamish and had the idea of replacing hanging with an electric chair as a means of human execution—all thanks to Edison's experiments born out of his determination to win this war of attrition with his erstwhile colleague.

But Edison could not overcome the nature of physics, and Tesla was ultimately proved right; his methods became the industry standard after a long struggle that left him a frus-

trated and lonely man. After falling out with Edison, Tesla
went off on his own, conducting ever more extraordinary experiments, and began regaling the media with increasingly
weird allegations. The base for this research was an experimental complex that he set up in Colorado Springs. It had a
200-foot-high tower topped with a massive copper sphere
and used huge coils to generate an unprecedented output of
up to 10 million volts of electricity. This resulted in energy
fields so strong that they ionized the surrounding air and
had devastating effects on the idyllic landscape.

Faucets in local homes poured out sparks instead of
water, and lightbulbs more than one hundred feet from
Tesla's tower glowed even when switched off because of the
high level of charge in the air. Horses in a nearby livery stable were electrocuted through their metal shoes as the
ground conducted this massive charge. Even local butter-
flies behaved as if drunk, the intense energy fields scram-
bling their sensitivity to the Earth's magnetism and
sheathing them in wreaths of sparks as they flew around
the fields in a frenzied state.

When asked what he was up to in Colorado, Tesla said
that he was attempting to manufacture lightning to be
used to power cities, but he was really delving into forces
deep within the structure of matter that he little understood. Powerful energy discharges were splitting apart the
atmosphere at a subatomic level and introducing side effects that science would not begin to understand for an-

other thirty years. As with most things, Tesla was decades ahead of his time.[15]

Tesla was undeterred by regular complaints from local farmers that he had killed their cattle with 135-foot-long lightning bolts (the accompanying thunder being reliably heard fifteen miles away); he was determined to give the world cheap power. Just one man believed in him, the rich banker J. Pierpont Morgan, who gave Tesla $150,000, a fortune at that time, to perfect a "radio wave generating tower" for a global communications network that would be cheaper and more powerful than any other. Morgan let Tesla erect a second massive generating tower on Long Island, New York. But once his benefactor grasped Tesla's plans to beam super bolts of power across the Atlantic amidst artificial lightning storms, he began to get cold feet. When it became evident that existing, already powerful energy companies would become obsolete overnight if Tesla had his way, his sponsor decided to cut off Tesla's funding and, quite literally, pull the plug on the project.

Tesla was already reeling from a huge catastrophe at his Colorado Springs base. A sudden discharge churned out by his lightning generator had fried the main electricity station for the area. The surge of energy pouring out blew the insulation and wrecked the place—an event that understandably added to the mounting protests and brought down the wrath of the giant Colorado Electric Company on the hapless inventor. The wrangling went on, but Tesla could not pay for the damage caused—having poured his now evaporating money into Long Island. Using an unpaid electricity bill of $180 as justification, the Colorado generating company seized Tesla's working tower, tore it down, and sold the lumber for firewood.

In the decade before moving to Colorado in 1899, Tesla made many intriguing observations during his power experiments. In that period he worked with the Institute of Electrical Engineers in New York and designed the first power station taking electricity from Niagara Falls and delivering it to the city. During several test runs strange balls of light formed around his towers *after* he had disconnected all his generators. These were a couple of inches in diameter and drifted freely around the area like clouds of energy. They then faded silently into nothingness within seconds. Tesla noted them, but did not understand what they were. His dream was to harness these balls of energy to light cities.

In March 1895, while conducting tests with his latest huge transformer in upper New York state, Tesla accidentally produced some enormously powerful rotating magnetic fields, which he said altered time and space in the immediate vicinity. It was not the first time some very odd things happened during his experiments. He often felt tingling sensations on his skin, suffered headaches, and experienced a sense of disorientation. These physical sensations seemed to be triggered by the energy fields that he was creating all around him. But he also noticed a weird feeling of timelessness, as if past and future were rendered without meaning. As he was ripping apart the atmosphere with his artificial lightning, Tesla pondered whether he was also breaking down the time barrier.

Tesla's most amazing experiment happened unexpectedly and scared the life out of him. Indeed, it may have fueled his later deterioration in mental health that is so evident from his life story, as it is only afterwards that he began speaking about death rays and Martians.

On March 13, 1895, he spoke with a reporter from the *New York Herald*. On meeting up as arranged at a café, the physicist was looking decidedly stunned and explained the reasons behind his trepidation to the curious journalist. That day, Tesla explained, he had narrowly avoided death when a massive ball with 3.5 million volts inside detached from his transformer, floated across the room, and struck him forcefully on the right shoulder. Needless to say, he had feared the worst, as he stood quite helpless in the path of this slowly advancing mass of energy. He only had time to utter a gentle cry of "oh!" before it was on top of him and turned reality upside down.[16]

Fortunately, Tesla's assistant was standing next to the switch and shut off the current as soon as he saw what was happening, probably saving the scientist's life. However, during this very close encounter Tesla experienced some bizarre sensations. He first suffered muscle paralysis, presumably as the energy scrambled the nerves in his body. Then, he says, he was taken out of time altogether and entered a magical realm where the true nature of time itself was revealed to him.

During this momentary experience Tesla saw that time was quite different outside our three-dimensional world. In his disassociated state, the physicist witnessed time as it really was in a universal sense—he saw it stretched around him forming another dimension with the recent past and the near future, all now equally accessible.

You could travel to this dimension, he said, just like you could walk into the next room. Movement through time was akin to movement through space. He believed that this proved that the energy he was creating was so intense that it was breaking apart local space and opening up pathways between times past and future. Once the gate was open, time could be seen crossing the cosmos like a series of highways. He eagerly wrote up his notes for future reference, hoping that he would gain the funding to build a time machine. But few people believed he was being serious, and he never got the chance.

Nikola Tesla has gone down in history as an eccentric inventor who went off the deep end after a promising start. Few scientists give much credence to his claims as the world's first chrononaut. But they should. Long after his death the march of modern physics has inch by inch revealed the nature of the universe, and much of it matches precisely what this Victorian researcher reputedly observed. If more of his peers had listened, time travel might already be an everyday reality.

1905
RELATIVELY SPEAKING

In the same year that Tesla may have accidentally broken the time barrier and H. G. Wells was imagining what a time machine might be like, a sixteen-year-old German boy was daydreaming, but his were not the usual wild thoughts of the average teenager. Instead the young Albert Einstein regarded the nature of time with such fascination that he worried about it day and night. Einstein believed that there were momentous discoveries to be made about these matters, and from his teen angst was born the proof that time travel was not merely possible, but that it could be an everyday occurrence.[17]

Einstein mulled over such questions with the innocence

of youth mixed with poor schooling. To compensate for his educational shortcomings he wrote imaginary stories inside his head, which unfolded like a film script built on real scientific issues. These scripts posed strange-sounding questions—such as what would happen if you were traveling at the speed of light and held a mirror in front of your face? Would the light reach this mirror and be reflected back to your eyes, especially given that the speed of the rays moving towards the glass should cancel out your own motion in the opposite direction? Einstein thought it absurd to envisage a mirror where no light rays could ever reach its surface, for time would then effectively stop if light was now unable to carry any images revealing ongoing events. Time standing still was unacceptable in his imaginings, and so Einstein concluded that the speed of light must be unlike every other known speed in the universe to somehow escape from the absurdity of this "thought experiment."

Einstein moved to Switzerland, became a clerk, and continued to wrestle with these cosmic questions, often when he should have been working. By age twenty-three he had conducted various other experiments in his head that he had neither the resources nor equipment to actually perform. But these limitations conveyed a big advantage because they let Einstein think way outside the box, which is exactly what was required to solve the persistent riddles that physics had endured since Galileo and Newton.

As the twentieth century dawned, science had calculated light speed accurately, realizing that it traveled at the same velocity as other types of radiation, such as electromagnetic (EM) rays, and could be transmitted as waves of energy. Yet this mounting evidence about the wave nature of light had been contradicted by the work of Albert Michelson and Edward Morley, two researchers based in Cleveland at what would become Case Western Reserve University. Indeed, Michelson was to be the first American physicist to receive the Nobel Prize, which was awarded for a lifetime exploring the speed of light.[18]

All waves of energy must ripple through something. Waves disturb the water in this way and sound moves through surrounding air so we can hear when these ripples push against our eardrum. So what does light ripple? The undiscovered substance was given a name—the ether—and it was assumed that space was full of it, rippling away as light waved through it. Unfortunately, nobody could find any trace of the ether.

Michelson and Morley had cleverly likened the ether to the wind. You cannot see it, but you can feel it as you rush through it. When you travel with the wind, your speed is boosted. If you travel against it, you are impeded. Every athlete or pilot must work with this effect. Michelson and Morley argued that the ether would alter the speed of light depending upon whether the Earth was moving with or against the direction of its flow. It seemed a very sensible suggestion.

To measure the effect, they designed a clever setup of light rays and mirrors that could be bounced back and forth so that one ray went with the flow and the other moved at right angles to it. Then these beams would be recombined. If light speed changed when passing at different directions to the ether, then the recombined rays would be out of phase with one another. However, no matter how often the scientists tried the experiment, there was no visible effect and there never has been, regardless of how much precision is brought to bear on the task. They even retried the experiment six months later to rule out the motion of the Earth when at different locations on its orbit around the sun. This made no difference either.

Their experiment revealed two huge problems. Firstly, if the speed of light could not be affected regardless of whether light was traveling with or against the Earth's travels through space, why then did our measurement of light not speed up or slow down? This made no sense unless the Earth was stationary, as it patently is not. But it was equally serious that there was no ether, nothing out there for light waves to ripple through. Einstein was determined to figure out why light was so extraordinarily different. His answer was the concept of relativity.

Galileo had already discovered relativity involving motion. And most of us experience this effect in our daily lives. Galileo noted, using a ship as an example, that if you stood on deck, you seemed at rest relative to your surroundings, though to an outsider you clearly seemed to be moving. This kind of relativity applies to all objects in motion and explains the momentary uncertainty we feel when a train next to the one we're in begins to move away. Are we the ones now traveling or is it the train on the next track?

Einstein took this as his starting point, imagining many experiments with trains, and later a series of experiments involving light beams. Imagine being on a train heading from New York to Washington. It is moving at 50 mph. If you walk down the cars in search of a snack, then you could go in one of two directions. Say you head at 5 mph towards the locomotive pulling the train. As far as you are concerned, you are moving at 5 mph, just as you would be if you walked in the opposite direction. That is because your speed is relative to the train you are in, and both you and the train are traveling as one. Yet, to a person standing in a station through which the train is rushing, the situation is quite different. If you are walking towards the locomotive then your 5 mph *adds* to the 50 mph of the train and the outsider sees you walk at 55 mph. If you go in the opposite direction then your speed is *deducted* from that of the train, because you are walking against its motion. Thus a person in the station sees you walking at 45 mph relative to the outside world. In other words, your speed can be 5 mph, 45 mph, or 55 mph, according to circumstance, and all these results are correct, because how we record speed is relative to the location of the observer.

By all logic, light speed ought to follow the same rules, but the Michelson-Morley experiment raised serious doubts about it and Einstein's thought experiment with a flying mirror convinced him that light cannot alter its speed regardless of relative motion. Shine a flashlight towards the engine on your train and you record the light rays traveling at 186,000 miles per second, just as expected. But it will travel at the same speed if you point the flashlight in the opposite direction. Unexpectedly, the light from the flashlight also travels at the same speed as measured relative to someone watching your actions while standing on the station platform. The speed of the train is irrelevant.

Einstein developed all of this into a theory known as Special Relativity that he submitted unannounced to a science journal in 1905. It was a massive shock to the old professors at the helm of physics to see an untrained, unknown upstart presenting results based on thought experiments. But the work was brilliant and sound. Time, said Einstein, is relative, but the speed of light stays fixed. The closer you move to the speed of light, the more the passage of time seems to stretch, or as scientists say, time gets "dilated."[19]

Time dilation is a legitimate form of time travel, and Einstein had become the first person in the world to prove through math that the time barrier could be broken. Although the rate of flow of time outside a moving object is unchanged, the faster you travel, the more it slows down for anyone or anything that is in motion. Because the moving object and its occupants experience less time pass-ing than the rest of the universe when the journey ends, they would have traveled into the future by virtue of having lived for less time than had passed in the world outside.

The faster you move, the stronger the effect becomes and the more you time travel into the future. But there was a big problem with such a journey. While it was very much a one-way trip, you reach tomorrow quicker than those who were not moving with you. But once you got there, you stayed there. Einstein's theory offered no ticket back.

Of course, the time dilation predicted by special relativ-ity is negligible, because we can travel only a fraction of the speed of light. The time jumps we experience do not exceed a tiny, tiny fraction of a second and are unde-tectable without the use of sophisticated recording devices. That is why nobody had spotted the problem in day-to-day life, even though Einstein had just shown that some-one running a marathon must time travel ever so slightly into the future relative to someone who merely walked the same route.

Light moves at 186,000 miles every second, but you must reach a substantial fraction of that velocity for time dilation to reach even a few seconds. To time travel ahead by days, weeks, or months you must fly at well over 99 per-cent of light speed. So Einstein's ideas, while amazing and mathematically sound and unquestionably proof of gen-uine time travel, seemed of no practical use given that we could barely imagine ever being able to go fast enough for them to matter.

Relativity theory produced many other extraordinary results. If you could travel at the speed of light, then your size would reduce to zero, your mass rise to infinity, and the effective passage of time vanish altogether. For this reason it is usually argued that Einstein proved that travel at the speed of light was physically impossible. Indeed, he originated the concept of an unbreakable light speed barrier.

Why? Because nothing can possess infinite mass. It would require infinite energy to move it and the universe cannot ever supply an infinite amount of anything. So relativity creates an impossible scenario where an object moving at light speed would be incapable of moving because it could not ever get sufficient energy to do so.

Nonetheless, some things do travel at light speed—light itself, for instance, self-evidently does! However, light is a wave of energy. It has no mass that needs to be accelerated. The restrictions apply only to material things like people or spaceships that would require infinite levels of energy to go that fast. Yet this restriction also has a curious result. From the relative framework of the beam of light everything would seem normal, but it would experience a period of timelessness and arrive everywhere in the universe instantaneously!

These conclusions hold great importance for time travel. We now know that moving at very high speeds will carry you into the future and slow your perception of time. Unfortunately, you could never stop time altogether by getting onto a spacecraft, or a time machine, and aiming to fly at light speed. But there is no similar restriction for energy waves.

This loophole seems to offer a startling new way of time travel. If one could beam energy through space at light speed, then very strange forms of time travel are allowed. Although seemingly absurd, the transfer of matter into energy became a reality not long after Einstein published his theories, and by 1945 had led to the atomic bomb and the nuclear age. So time traveling by way of transforming matter into energy is not as ridiculous as this might seem.

These bizarre implications emerging from relativity theory shocked many scientists. Einstein seemed to have given birth to various possible ways to travel through time that mankind might sooner or later find a way to exploit. If perfected, they would bring to the fore all those paradoxes that time travel should invoke. Small wonder then that physicists soon had doubts about how right Einstein was.

Relativity itself did not seem to be in error—try as they might, physicists could find nothing wrong with Einstein's thinking. But time travel was unwelcome, so, many hoped, it would eventually fall when his theory could be tested. Of course, right or wrong, Einstein had not solved the mystery of light, which had been one of his primary intentions, although he helped show why it seemed to be so confusing. In a letter to his friend Michael Besso in 1951, he commented: "No-one really knows exactly what light is. All we know is that it can be both a particle and a wave. And that's the paradox. How many things do you know of that can be two different things at exactly the same time?"

Despite the genius of his ideas, Einstein never won a Nobel Prize for relativity. Relativity seemed so removed from everyday experience that it received little attention. But beginning with the year this epoch-turning theory appeared, would-be time travelers gained their first big ally.

Time travel became scientifically credible in 1905. It marked the moment when the prospect of a time machine turned from a fantasy to a possible reality.

1919
BENDING SPACE AND TIME

Special Relativity is called "special" because it relates to a limited set of circumstances and omits various factors that would always apply in any real situation. Unwilling to let his theory rest incomplete, the forward-thinking Einstein set out to develop a more generally applicable theory to include such things as the universally present force of gravity. But in doing so he would merely introduce yet more nightmares for those physicists still reeling from the time travel implications of his original research. Einstein's new work would fire the starting gun in the race to build a time machine.

For all practical purposes, even when dealing with

speeds of many thousands of miles per hour, the laws of motion defined by Newton still worked perfectly well. Einstein's rules of relativity simply modified Newton's equations; they did not replace them. However, our modern technology does sometimes need to consider relativity when great accuracy is called for. We must factor the calculations into global positioning systems, for instance, because these use signals bounced off satellites in space and relativity alters the measurements just enough to give a slightly false reading otherwise. Atomic clocks also need to be accurate to within tiny fractions of a second and are compromised even when the clock is traveling at modest speeds. Without factoring relativity into such technologies, we would find nagging errors creeping into their operation that would otherwise be unexplained.

In 1915 Einstein published the results of more thought experiments revealing his general theory. He said that space and time must be considered as one, in order to define the nature of gravity, and he called it space-time.[20] This force was not, as physics had assumed since Newton famously saw an apple fall off a tree, an attraction between one body and another—with the heavy mass of the Earth pulling the lighter mass of the apple into the ground. It was all the result of a depression in space-time.

To envisage how this works, think of a sheet of rubber stretched taut over a large area with pegs staking out each corner. This sheet represents space-time. Consider an object moving on a path through the universe—which in this illustration means that it must roll across this sheet of tautened rubber. A baseball, for instance, will slightly depress the rubber as it rolls across the sheet, which will consequently stretch beneath it. But if the ball is much heavier, made out of lead for example, then it will bend the rubber much further, creating an even deeper well of depressed rubber formed beneath its weight. This deep well will always have curved surroundings regardless of the mass of the object creating the depression.

All objects create gravity by bending space-time in this way. The greater the mass, the deeper the well it creates, and the steeper the curves on its sides. Because these wells have rounded edges, anything moving through space-time will be forced into a circular path to skirt the edges of the well, resulting in the planetary orbits that we see throughout the universe. Any orbiting planet is forced to alter its motion through space, but is also distorted in its course through time. The more intense the gravity well, the greater the effects that result, sucking objects down like water spiraling into a plughole. Greater and greater speed may be needed to compensate for this downward motion that is enforced by the gravity well. But a sufficiently deep well might require an object to move so fast that it would have to exceed light speed in order to avoid getting sucked in. That, of course, is impossible. So in these circumstances the object might be trapped forever in the well.

The depth of the resulting gravity well increases when massive objects pass through a region of the universe. Anything that is particularly large will warp space-time to a serious degree. The greater the mass, the more space-time will bend, and the more intense its effects, which will be felt by anyone coming into its vicinity—for instance, a hapless space traveler passing nearby. But Einstein showed that space and time are inextricably linked, so massive objects will create enormous distortions on both time and space. Any passing object will have to navigate these changes to space and time as it crosses a region affected by this huge gravity well. Time distortion and time travel will result because both space and time will differ greatly from the norm and the passing traveler will experience sudden shifts in location and alterations to the flow of time that will manifest as traveling through both space and time. Needless to say, this prospect did not sit well with some physicists.

This amazing new picture of the universe also has a bit of a synchronistic twist. Einstein forces us to conclude that space-time is bent into wells of high gravity. These high gravity (HG) wells thus provide the key to time travel—a concept first defined in fiction twenty years earlier by a rather different H. G.—science fiction novelist Herbert George Wells!

Relativity seemed impossible to prove, but in 1911 Einstein finally proposed how to demonstrate some of its key components experimentally, based on the concept that any heavy mass will create a high gravity well and bend space-time. During a total eclipse of the sun, the sky darkens as the moon is precisely positioned between the sun and the Earth. In those few minutes of totality, stars very close to the sun that are in line of sight with the Earth become visible even in the middle of the day. These are the very stars whose light rays will be most affected by the huge gravity well created by the sun's mass. The light rays will be bent around this well and will slightly alter their position as seen from Earth. Einstein predicted precisely how much warping of this light should occur according to relativity theory. If the light from these stars did bend during an eclipse, then relativity would be proven right; if not, then his theory was wrong.

The renowned British physicist Sir Arthur Eddington agreed to take on this challenge. On May 29, 1919, two expeditions were mounted to view an eclipse and photograph stars very close to the sun at the moment of totality. These stars were first photographed six months earlier (when the sun was in its opposite position in the Earth's orbit and the sun's gravity well had the least possible effect). These measurements were used as the baseline from which any distortion could be detected. Relativity predicted that the mass of the sun would curve light from these stars through the tiny but measurable distance of 1.75 seconds of arc. Results coordinated by Eddington averaged 1.79 seconds of arc, an almost exact match given the level of accuracy for the instruments of the day. The outcome was revealed to a packed meeting of the Royal Society in London six months after the eclipse. Scientists hailed this experiment as the most important development in physics in centuries because it verified the stunning ideas behind relativity. The next day the *London Times* carried headlines alleging that Newton had been overthrown! This is not exactly correct, but it showed the momentous nature of Einstein's work.[21]

Einstein was now a media celebrity and spent the last thirty-six years of his life seeking to unite the worlds of the very small (quantum physics) with the very large (relativity) because it was immediately apparent to him that they seemed incompatible. Scientists are still searching for this "unified theory" that will link these worlds (the TOE— Theory of Everything—as it is known). Experiment after experiment has proven that Einstein's research in both these fields stands up to all scrutiny and that the very small and very large must somehow be united in order for us to fully comprehend how the universe works.

But there was a far more worrying consequence to the proof of relativity, as far as many scientists were concerned. Time and space could be distorted by a heavy mass interfering with the passage of electromagnetic radiation such as light. If this happened in nature, then perhaps it could be reproduced artificially in the laboratory. Suddenly there was a creeping realization that relativity could be manipulated by human ingenuity and once we succeeded in creating artificial gravity wells the prize could prove to be immense.

Distort time and you open the barriers that prevent us from traveling to the future or the past. Einstein had shown that it was possible to build a time machine without reaching the speed of light.

1926

OTHER DIMENSIONS

The first person to respond to Einstein's challenge and construct a working model of a time machine was Scottish engineer John Logie Baird. Yet nobody, including Baird himself, realized that he had done so at the time. This was in 1926, a year that was to prove very significant in the race to break the time barrier.

As the inventor of TV transmission, Baird would give Britain a crucial start to let the BBC transmit the world's first regular television schedules in the 1930s. But his interest in exploring light waves and the mysteries of Einstein's space-time mirrored many other scientists of the day who were eager to see where relativity might lead in practical terms.

While working on the technology that would provide his immortality, Baird designed a camera that would reveal some of the unseen energy waves that Einstein's idea of space-time suggested were awaiting discovery. In his memoirs for 1926, Baird described the simple design of his new device. It blocked out light from an electric bulb and let only infrared radiation emerge. He cleverly realized that this camera could also detect body heat, which was radiation from the same part of the spectrum.

He tested the camera by placing Sir Oliver Lodge, the renowned physicist and pioneer particle researcher, in a totally dark room. No ordinary camera, or human eye, could reveal this elderly gentleman's presence. But using his "night vision camera" Baird created an image of the distinguished scientist on the screen, noting that his long white beard showed up particularly well. Such an exposition of invisible dimensions undetectable by our normal senses was a demonstration of our rapidly unfolding knowledge of the universe. But why this was also a time machine was barely obvious in the 1920s.

Baird rather oddly chose to use his invention in an eccentric manner. As the talk of the day was about how our reality might contain whole universes buried out of sight within Einstein's space-time, the idea was to use artificial aids like this camera to look for them. Perhaps his night vision device was exposing the first signs of other worlds that the eye could never see. Moreover, he mused, perhaps one of these other dimensions was where dead people lived after shuffling off their mortal coil.

This seems a bizarre step for a scientist to take, but his motives were otherwise orthodox. In fact, he was commendably keen to destroy such thinking, and not reinforce the idea of life after death. So he practiced using his camera at séances to try to expose fake mediums and disprove the then popular belief in an unseen spatial dimension we call the afterlife. The plan was to catch any charlatans who moved objects while the room was in complete darkness— as Baird was sure that many did—thus appearing to those they duped to have commandeered the spirits from another realm.

This amazing new viewer was the key to exposing this trickery and would hopefully shock the perpetrators, unaware of its unusual capabilities. And Baird did just that on several occasions. However, he was not to emerge from the tests completely skeptical, and he reported in his memoirs that there were times during his work when his device seemed to provide evidence that a few mediums were truly interacting with other dimensions. "I witnessed some very startling phenomena," he advised, "under circumstances which make trickery out of the question . . . I am convinced that discoveries of far reaching importance remain waiting along these shadowy and discredited paths."[22]

Nor was he alone. Even the then elderly Thomas Edison told *Scientific American* that he was convinced that you could construct a device to probe into the invisible sphere where people went after physical death. He tried, but was confronted by his own mortality before he was able to succeed. Yet, according to Baird, even death seems not to have stopped Edison from continuing the quest! The Scottish inventor describes one extraordinary séance that he attended in Wimbledon in 1931 soon after Edison passed away. Using his camera he detected a strange purple cloud that appeared in the room, and signals tapped out in Morse code seemed to come from within this floating mass. Was this more mumbo jumbo acted out by a phony medium? Baird thought otherwise. These messages, he reported, told the scientist that Edison was passing on research tips to his successor from his new abode in another dimension. Baird said that the deceased Edison "had been experimenting with noctovision"—as he called this camera—in order to try to break the barriers of space-time.

Edison professed to be "convinced that Baird's device would in time prove of great use." That was indeed a far-sighted prediction, although the success has probably come in a way that was not anticipated. Today what seemed in the 1920s just like a bit of tomfoolery is regularly being used to search for people trapped under buildings after earthquakes and other natural disasters. The buried person's body heat gives off infrared radiation and the modern equivalent of this camera makes an outline of them visible beneath mounds of debris. Baird had constructed a machine that revealed aspects of reality outside those realms accessible to the human senses. And later it would prove its worth as a time machine.

In 1927, shortly after Baird created his camera, the great German physicist Werner Heisenberg proposed what he called the "uncertainty principle" that underpins the nature of physical reality. It states that "The more precisely the position is determined, the less precisely the momentum is known in this instant, and vice versa." In a sense this meant that reality remains "uncertain" until we observe it. This is because the act of observation changes either the position or momentum of the particles inside any atoms that we inspect. Looking at something alters what we see. So the universe cannot ever be observed with completeness or total accuracy. Why? Because perception is in some sense always an act of creation. Such a concept was as revolutionary as relativity and just as shocking to science. The implications for time travel were also profound. Did it mean that we could alter the past simply by observing it?

During this remarkable 12-month period, a colleague of Heisenberg's, Oskar Klein, detailed another theory that would add momentum to this strange concept—that we are the architects of and not mere participants in a universe full of unseen realities. Klein sought to link this new image of the reality inside the atom with Einstein's theories about space and time on a cosmic scale. In doing so he showed that it was impossible to build a bridge between the big and the small without conceiving of extra dimensions other than those with which we are all familiar.[23] He argued that physics would never be able to comprehend the nature of the universe without accepting that such extra dimensions existed even if they were invisible to us.

Think of it this way. To direct someone to any place you need four pieces of information. Starting from a point in, say, New York City, you can use words to allow someone to find a building in that town by describing two spatial dimensions, north-south and east-west. They could then probably find the building, but you would also have to specify the position in an up-down direction in order to take them to the precise apartment. The coordinates provided by these three dimensions define the location of any place, and we all use such thinking every day. But you also need to know at what point in time you intend someone to visit. Otherwise they might go to the correct place but long before you arrive to meet them or even before the apartment has been built. Indeed, because everything in the universe moves relative to everything else, even a stationary building is moving along with our planet through space-time. Consequently, to properly define any position it is always at least partly about time, not just space.

You might think that four dimensions, with time loosely defined as the fourth, would be sufficient. But Klein proved that it was not. In order to link the quantum world with the grander picture of space-time, a further spatial dimension—

a fifth—was required. He came to this conclusion because experiments with subatomic particles were revealing some extraordinary occurrences. In the space within matter particles appeared, collided, disappeared, and passed through one another like ghosts and yet could never be pinned down as to where they were actually located. Time travel makes sense against this backdrop because if it is impossible to say where something is positioned at any given moment, then it is also impossible to say when it might pass any point in space. This suggested that time was fluid and travel through it perfectly possible within nature.

Klein suggested that only another dimension of space that we were incapable of viewing could explain these crazy occurrences. Particles appearing, disappearing, moving across space, and reappearing somewhere else in time may seem like magic, but they only look odd because the particles are moving normally but doing so through a dimension that we cannot see.

Klein was also able to explain why we cannot readily detect this fifth dimension. It is microscopically small—far smaller even than atoms or the particles that form the space that our senses can detect. We have searched for smaller and smaller constituent particles ever since the atom was split into protons, neutrons, and electrons that are its essential makeup. But the subatomic particles being identified were so minute that there was no way to directly observe any of them. Klein preempted today's opinion that there could be whole realms of experience formed by the tiniest things within the universe that formulate whole dimensions of space that are far beyond our capacity to observe. When energy waves and particles move through this fifth dimension they vanish and then suddenly reappear in another place in our dimensional space. In reality, from Klein's perspective, they just nip down an inter-dimensional rabbit hole and move via this hidden realm to another spot in our universe. But by moving across space, bypassing the rules that exist within the space we inhabit, they cheat time as well. In fact, these particles must time travel relative to us.

Klein argued that this fifth dimension must be very small and curled up on itself in order to be invisible, and that it was probably, in fact, embedded into the structure of normal space like tiny air bubbles unseen inside a candy bar. At first Einstein found this theory patently absurd, and said so, but the more he studied it, the more he came to agree with Klein that relativistic space-time supported the concept. Indeed, by the 1930s Einstein was an eager researcher into the possible consequences of an unseen fifth dimension.

But why was all of this talk about invisible dimensions to prove so relevant to time travel?

Some years after Baird's death, military scientists at Eglin USAF base in Florida conducted an experiment using a second-generation version of Baird's noctovision camera. Mounted aboard a high-flying aircraft, the camera recorded infrared energy from a parking lot thousands of feet below.

The parking lot was completely empty at the time, but inexplicably the photographs showed the ghostly images of whole lines of parked cars, as if these vehicles were present but invisible to the human eye.

What were these cars? Were they in another dimension? In fact, the truth was stranger than that. The cars were real but had left the parking lot some time before the flight passed overhead. This camera was acting as a time viewer, revealing images of vehicles that had existed in this space at an earlier point in time. The energy emissions detectable from the cars before they had departed could be used to re-create the residual outlines of their body shapes. A sophisticated infrared viewer could recombine the signals into pictures showing the cars exactly as they were when parked in the lot. This device effectively brings the past back to life by using those hidden energy emissions that our senses otherwise cannot detect.

Space and time are intimately related, as Einstein had proved. Any camera detecting things from other dimensions through space could be turned into a camera to view other times. Einstein also came to realize that extra dimensions beyond Klein's fifth might be needed to explain reality. Indeed, the number of dimensions that may be out there beyond our perception seems to grow every few years as scientists recognize their necessity. At present some theories argue that we may need as many as twelve dimensions to explain reality!

Perhaps the paradoxes of time travel that so terrify some scientists are illusions that vanish if we see the events from an extra dimensional perspective. This realization spurred many formerly skeptical physicists to dig deeper into these possibilities.

The quest to build a time machine received its greatest boost with these events of 1926. People now realized that time travel need not be about speed at all. It could just be a matter of designing a vehicle that would cross unseen dimensions. Time travel would merely be the price you pay for buying a ticket.

1935

BRIDGES ACROSS TIME

In 1927 a gathering of scientists discussed the weird theories and experiments that had tumbled out in profusion since relativity had been discovered. Anyone who was anyone, from Einstein to Heisenberg, took part. This meeting concluded that nobody really knew what the heck was going on, but that these theories worked. This gave the green light for research into the fifth, and sixth, dimensions. It legitimized the hunt for a practical time machine.

One thing this illustrious meeting did agree upon was the validity of the uncertainty principle. All energy and matter exists in a state of flux until some experimenter or measuring device decides to observe it. Until then it possesses just virtual reality, mostly as a series of statistical probabilities. Only when measured in some way did statistics solidify into actual reality.

Einstein was less than enamored with what he saw as a thoroughly silly concept. As he pointed out to his colleagues: "Are we thereby now to suppose that the moon ceases to exist if we stop watching it?" He could not tolerate what he also laughed off as God playing dice with reality.

To fight back, he joined forces with colleague Nathan Rosen to find a definitive test to prove that our view of subatomic physics was wrong. Using a thought experiment, Einstein and Rosen imagined twin particles that split apart during a subatomic event and flew off until they were light years from each other. According to quantum theory, these particles could now communicate with one another across the vastness of space and do so faster than the speed of light. This communication resembled telepathy. Needless to say, there was no way Einstein could accept such an outcome, so he hoped that the patent absurdity of his experiment proved that quantum mechanics must be in error. But Einstein made little headway in convincing the majority of physicists because they could only argue that the theory otherwise worked quite well.

Undaunted, Einstein and Rosen went back to looking into the issue of hidden dimensions embedded within space. They had good cause to do so. Just as Einstein was skeptical of quantum theory, there was a growing mood that somehow relativity must be incomplete, because of its time travel consequences and worrying temporal paradoxes. To overcome this drawback Einstein and Rosen investigated how motion through unseen embedded dimensions might allow what looked like time travel to our eyes without violating any sacred rules, especially faster-than-light travel.

The two researchers then made a stunning proposal. They imagined regions of space-time that acted like a bridge through those hidden dimensions and in the process linked two different parts of the universe that we could see. It was like taking two buildings, one on either side of the highway, placing a ladder from one roof to the other, and using it to ford the gap. The Einstein-Rosen Bridge, as it was known, simply crossed a similar gap through unseen dimensions.

If you need to go from one building rooftop to another, the normal way would be to climb down to ground level, then enter the other building and climb up to the roof by the stairs or use an elevator. This is a long journey and would take some time to complete, yet it seems to be the only obvious way to make the trip. But if you dared to cross the ladder that bridges the rooftops, then you could travel from one building to the other much quicker because it vastly reduced the distance you needed to travel.

In exactly this same way Einstein and Rosen envisaged bridges through the unseen dimensions of space that could reduce any journey by taking a shortcut between two points.

The journey would take far less time than by traveling what seems the only direct route. If subatomic particles could use this cosmic ladder, they would seem to move through great distances at faster-than-light speeds, which would appear to us as time travel. Yet faster-than-light travel contradicts relativity theory, hence the suspicion of physicists that relativity was somehow flawed. But if the particles had crossed a bridge through an unseen dimension, then these particles would not need to travel so far or so fast nor be breaking the laws of physics.

Einstein-Rosen bridges are now more often called "wormholes," a term that better illustrates how they work. If you think of the curved nature of space-time as the surface of an apple, then you can mark two spots on opposite sides of the fruit. If you measure the distance between these spots by the only obvious route (halfway around the circumference) you may find that in a typical apple this distance is, say, five inches. However, there is another way to journey across the apple—the path a worm might take by burrowing through the core. In this way a more direct trip is possible that may measure only three inches.

Wormholes in space-time enable the same sort of shortcut a worm uses, shunning the surface dimensions in favor of burrowing through the apple. We can readily visualize this process in terms of two and three dimensions because our senses are designed to operate in three dimensions, but our senses are not equipped to see the fourth or fifth dimensions through which wormholes might cross space. That's why they look like magic to us.

Wormholes are the simplest way to achieve time travel using a spacecraft. When Einstein and Rosen discovered the concept, it was just a theory, but plausible enough to instantly demolish those skeptical arguments that relativity required faster-than-light speeds to allow time travel.[24]

Remarkably, Einstein's perceptive theory was preempted by a 1934 science fiction novel by John Campbell entitled *The Mightiest Machine*. In the story the author overcomes the problem of having a spaceship cross the enormous distances between planets by imagining a way to compress these distances by folding space and squeezing it together. It would be like taking a long strip of paper with the departure and destination planets marked at each end, then crumpling this into a ball so the two points on the strip would now be next to each other—meaning that any trip between them could be cut short.

Campbell called this hidden dimension that reduced space "hyperspace"—a term widely used by science fiction ever since. Great minds think alike, as they say. Einstein and Rosen reached the idea of bridges or wormholes for different reasons (to try to salvage relativity against its critics). Campbell had only a loophole in a plot to fill in order to tell a good story. But both may have hit upon nature's own solution for interstellar travel and journeys through time.

1937
TRAVEL INTO
THE FUTURE

Eddington may have provided spectacular proof of relativity during the 1919 eclipse, but that experiment only demonstrated that Einstein's concept of space-time was accurate. Nagging doubts remained over the issue of time dilation. It seemed against all common sense to imagine that if you ran to catch a bus rather than walked the same distance, it caused you to time travel into the future. But Einstein insisted that even though we cannot measure the tiny amounts of time traveling that we all do in this way, the principle was nevertheless sound. If we could travel at very great speeds, the effect would be obvious, and ultimately we would be seen to time travel in a meaningful fashion.

Of course, we need not worry about changing the past. Time dilation involves faster travel into the future, and as that is unformed and open to alteration, there might not be the same damaging paradoxes as time travel into the past entails. But time travel, however it's done, has always been too much for some scientists to even consider.

The opportunity to test this troublesome concept of time dilation came in 1937. Advances in the understanding of the subatomic world had now shown that there were many constituents of matter, each microscopically small. The electrons, protons, and neutrons that were revealed as the main components of an atom were now well established but many exceedingly tiny particles that helped to formulate the bigger ones also existed, mostly for just a very short time. These particles were created out of the waves of energy that filled the universe as they interacted with one another. They formed, broke apart, and gave birth to other particles and this hugely complicated dance involving trillions of different units of matter went on amidst a seething mass of energy waves.[25]

We cannot directly observe these events happening at a subatomic level but can infer them through what are known as quantum fluctuations, where blips in the statistical rules governing these energy waves coalesce to form matter from this maelstrom. What's produced are called "virtual particles," since they are not real in any usual sense, but not imaginary either! Looked at another way, the space within atoms, which seems empty to our senses, is really a simmering mass of energy boiling off frenziedly into numerous bits of matter. Once actualized, these particles meet each other and swiftly annihilate, thereby appearing to create matter out of energy and then generating energy from the resulting matter collisions. We lack the ability to see virtual particles in situ, but they are real in every other sense—just hidden from our perception.

One of these newly found particles was identified thirty years after Einstein had discovered time dilation. It is called a muon. These are very tiny, weigh next to nothing, and so can be made to move extremely fast. But these particles travel only at modest speeds when they are formed during physics experiments and they exist for just 2.2 millionths of a second, after which they self-destruct into various other particles, mainly electrons and neutrinos.

However, muons also reach the Earth's surface in a natural form. They come here in showers of cosmic rays that fall like unseen rain onto our world from the depths of space, doing so alongside many other particles—some of which pass right through our planet as if it were not there! The muons arriving in cosmic rays have been accelerated through the universe at fantastic rates and are traveling very close to the speed of light by the time they arrive on Earth. These naturally occurring muons behave quite differently from their slower-moving cousins that can be created in a laboratory. The biggest difference is that they survive much longer than the slow-moving variety.

BREAKING THE TIME BARRIER

If you calculate the effects theorized by special relativity, a longer life is exactly what time dilation predicts. The faster something moves, the less time will pass within its frame of reference. So from our perspective a fast-moving muon has experienced only part of its lifetime thanks to time dilation, while the full 2.2 millionths of a second life span passes by in the universe. The muon has time traveled by those moments into the future. Trivial to us, but not to such a short-lived particle.

Today it is possible to accelerate muons in a lab to ever more rapid speeds and to measure the increased length of their existence as we do so. Moving at 99.5 percent of light speed, muons survive for 10 times as long as they do when moving sedately. These numbers are precisely in line with time dilation theory.

If you equate a muon to a human being and thereby imagine that its 2.2 millionths of a second is equivalent to a person's 100-year life span then we can see how humans would experience time travel by moving at near light speed. At everyday speeds, time does not move forward fast enough for us to notice. However, if we traveled at 99.5 percent of light speed, just like these muons do, then a journey that lasts ten years in the outside world would only take one year for us. On reaching any destination at that speed we would be one year older, but it would be nine years into the future of the rest of the universe because the full ten years would have passed in the world outside. This is what life is like for a muon.

The extended lifetimes of these muons not only prove the concept of time dilation, but they are direct evidence for time travel happening in nature. Subatomic particles like these voyage into the future all across the universe. It is not a theoretical phenomenon but a demonstrated fact. This experiment with muons was the first to so readily document real time travel in action. In modern times we have been able to experiment further. Jet aircraft and rockets allow us to travel at thousands of miles per hour, just a fraction of light speed but enough to create now measurable time dilation effects in everyday life.

The alteration of time measured inside fast-moving objects was revealed in a clever way in 1971 when two physicists from Washington, Joe Hafele and Richard Keating, used highly sensitive clocks to measure the passage of time on a jet. These atomic clocks can record very small fractions of a second to great accuracy using the motion of particles inside cesium atoms, whose periodicity is well known. For this revolutionary experiment the scientists took the clock on a plane flying around the Earth in an easterly direction. This meant that the speed with which the Earth was rotating was added to the speed of the aircraft, making the clock on the plane move faster by several hundred miles per hour than did its twin left back in the lab. The clock left behind was not moving, except like everything else on the planet along with the motion of the Earth. Relativity argues that the clock on the jet should experience a slowing down of time, thus leading to the time dilation effect for anyone traveling with it. Although the aircraft was moving at much, much less than the speed of light, the results were still measurable. The atomic clock on the jet did age ever so slightly less than the one left in the lab (in fact by less than a billionth of a second). What this shows is that every time we travel very fast through space, we also travel into the future, exactly as Einstein had argued. The effect is real and its consequences inescapable. Unfortunately aircraft speeds are trivial in cosmic terms and even to time travel by a millionth of a second would require a velocity way beyond the fastest jet.

Time dilation has since been measured on board spacecraft, and the faster the occupant moves through the universe the more they travel into the future. Even at the highest speeds that we can achieve in space such time traveling still amounts to barely a few milliseconds. But Russian cosmonauts—some of whom have spent many months circling the Earth at tens of thousands of miles per hour living on a space station—have set new records for human time traveling. They are effectively living those few milliseconds in the future compared to the rest of us.

Einstein is truly the father of time travel into the future—having revealed a cosmos that is literally full of tiny subatomic chrononauts breaking the time barrier with apparent ease. Once this fact was proven beyond doubt by experimental evidence it inspired a number of young scientists to ask themselves one burning question. If these particles can time travel as part of the natural order of things, then surely human beings can be empowered to follow in their tracks?

TIME TRAVEL INTO THE PAST

1949

After World War II, the rockets the Nazis had built to rain terror upon London were swiftly repatriated to the United States, where these rocketeers kick-started the space program. Meanwhile, in a rubble-strewn Europe, French aerospace engineer Emile Drouet launched what he hoped would be a new kind of rocket ship. This one would not vent fire on a distant city nor reach for the sky climbing towards the stars. Instead it would travel through time.

So convinced was he that his craft could visit the past based on the work of Einstein and the relativity theory about the nature of space-time that Drouet built a scale model hoping to attract investors. In 1946 he put this on display in the town of Vigneux-sur-Seine.[26]

Drouet argued that the Earth moves through space in specific paths as shown by Einstein's math. We revolve around the sun at approximately 150,000 mph, but we do not notice this phenomenal speed because we are all traveling together. Relatively speaking, we appear stationary within our own environment, but that is an illusion. The sun is also moving at high speed towards a distant point in the galaxy. This means that if we think in relative terms, the Earth is voyaging through space in a rather more complicated series of motions, not just a simple orbital path.

To calculate the full picture we must add both the rotation of the planet and its journey across the universe tagging along with the sun. Indeed, it is more appropriate to think of this complete path *as creating a spiral through space-time*. Physicists who chart space-time routinely make similar, although often more complex, calculations. They produce mathematical diagrams that show in pictorial form how these various motions combine and the course that anything that travels through space-time will take.

Drouet wanted to use his rocket in such a way that it would match the speed of the Earth through the universe and travel on a path directly opposed to its tight spiral through the cosmos. In his opinion time was nonexistent except in the sense that it results from motion through space, so the Frenchman rather dubiously theorized that any ship routed in this fashion would travel backwards in time and find itself moving one year back for every spiral of the Earth's path that it retraced forward in time.

In other words, if the Earth has transcribed a spiral through space-time with ten loops occurring across ten years of time, Drouet imagined that by retracing that spiral through all ten loops you would backtrack in time and arrive ten years into the past. However, space-time curves are not so simple as straightforward spirals and such a concept operates only at a very basic level of plausibility. Drouet's time ship would not have worked in reality.

To be fair to Drouet, he had not envisaged carrying a human pilot but imagined that if this ship could be precisely calibrated it could use cameras and similar equipment to capture images from the past and relay them automatically back to the present. That might obviate the need for any daring traveler, especially as they would have bought a one-way ticket to yesterday. The only way to fly home would be the normal method—living their lives one year at a time. Of course, if the time rocket was sent back 200 years on its temporal reconnaissance mission then this would not be much comfort to any pilot hoping to get home to his loved ones. Any film taken in the past would not face the same dilemma. It could be left in a predetermined location to be retrieved by the rocket scientists from where it would have remained for those 200 years since completing a journey that may in our terms have only begun the day before!

Drouet had a kernel of an idea, which makes sense if one sees the spirals not as physical motions that we can mimic but as mathematical patterns in space-time. If you fly through space in that way, time travel might be feasible under certain conditions, but it would need a device far more sophisticated than Drouet's rocket and it would certainly have to travel much faster than 150,000 mph.

Drouet was never able to persuade anyone to back his brave plan, but at least he showed that the era of the time traveling engineer or scientist had well and truly arrived. He also foresaw twenty-first-century space probes using automatic cameras—an insight for which he certainly deserves credit given that in 1946 the idea of using rockets for spaceflight was still largely science fiction. Drouet's was the first brave plan to use modern science to design a time machine, but it would be the first of many. The media loved Drouet but scientists, of course, largely ignored him.

Einstein may have been the father of time travel, certainly in the form of journeys into the future. But it was one of his closest colleagues who would give birth to the nightmare scenario that physicists had long dreaded—he found a way to time travel into the past. In 1949, even as Drouet was still seeking finance to build a full-scale version of his time rocket, mathematician Kurt Godel horrified scientists with a model of the universe that allowed trips into yesterday without violating any known laws of physics. It was, in some ways, a mathematically sophisticated version of the French engineer's time rocket.

Godel worked with Einstein at Princeton and was highly regarded. Though his calculations were swiftly proven correct, a fortuitous escape clause remained for all the shell-shocked scientists. Godel had found that journeys into the past were possible, but only if the universe was rotating. In such a version of the cosmos, you could ride into yesterday aboard a rocketship that traversed paths not unlike Drouet's spirals, though more mathematically complex. Godel's "closed time-like curves" emerged naturally from a rotating universe. In such a universe, by flying fast, but below light speed, you could move in a circular path and visit the past and eventually return home to the present.

Although Godel's math was perfect and others soon confirmed that time travel into yesterday would indeed be a simple matter of steering the right flight path through such a cosmos, the problem was that our universe does not rotate as Godel's theoretical one required. By 1949 astronomy was sufficiently advanced to have allowed us to measure how the stars moved apart from one another and to calculate some of their basic parameters. These established without much doubt that our universe has a form that does not ordinarily result in the closed time-like curves that Godel had found.

Godel had done serious damage to long-standing theories about time and the possibility of traveling through it.

By defining a viable way to travel into the past without violating any known rules of physics, he had shattered decades of complacency. In effect, by opening the door at last to two-way travel Godel had single-handedly revived the race to build a time machine.[27]

1963
BLACK HOLES

It must have been very exciting to be a physics student at Cambridge University in 1963, for the physics department was at the center of the biggest debates of the century. Many of these concerned the nature of space and time and were to play a major part in helping to break the time barrier. You were just as likely to see TV crews and reporters milling about as you were academics strolling down the hallways of this ancient establishment. And the reason for their attention was often one man—a rough, tough-speaking northerner who was at home in the lecture theater and relished addressing the public. His fans adored how he would speak his mind and loved to express big ideas in simple ways through science fiction.

Fred Hoyle was the British cosmologist who had conjured up the name for the "Big Bang" theory in the hopes of ridiculing a concept that he detested. But it was his own theory of a stable universe that would be doomed by events—and time travel had a major role in its downfall.

Hoyle had great success with his classic 1957 story, *The Black Cloud*, which describes an intelligent gaseous organism roaming the universe and stimulating barren worlds with the seeds of life. This was the imaginative basis for his later real theory about how epidemics might spread when the Earth's orbit passes through the chemical tails of wandering comets. He suggested that these comets sow both life and disease like a cosmic farmer onto hapless worlds in their path. Such a strange idea was at first rejected, but became more credible with the growth of knowledge about the universe as we probed the depths of space. Some very unexpected objects were found out there—including strange lighthouse-like beacons pouring massive energy through surrounding gas clouds. These were so odd that some thought them to be messages from aliens, but they were really huge pulsating quasi-stellar (star-like) objects—hence being known as "quasars." Such findings, which had been suggested by Hoyle's delightful novel, inspired him to refocus the concept as a factual hypothesis that has yet to be disproved.[28]

Hoyle's figure dominated the Cambridge campus when a young man named Stephen Hawking was just starting out on his own epic career. Hawking is perhaps the only modern physicist to be well known outside of science, even to the extent of having guest roles in both *Star Trek* and *The Simpsons!* Hawking would become a vital player in the time travel story, but was fired up in those early days by Hoyle's arguments against the Big Bang. Hawking at the time had a great ally in the form of a younger member of the teaching staff, now Oxford don, knighted for his services to science—Sir Roger Penrose.

Penrose studied the massive objects then being discovered in deep space and their effects on the rules of space-time. He discovered that they must produce enormous gravity wells and that these would possess remarkable properties with profound implications for time travel.

Penrose's work followed that of the German physicist Karl Schwarzschild, a tragic figure who in 1916 made some calculations on whatever paper was at hand during rare moments of inactivity on the Russian front. He sent the news of his amazing discovery from the battle lines back to Einstein but never lived to see its importance, succumbing to a disease picked up in the trenches.

What Schwarzschild had discovered was that any star with a sufficiently concentrated mass would create a gravity well that was so intense it would start to collapse in on itself. Any person standing on its surface would be crushed. The gravity well created at the point where your feet stood would be vastly deeper than the one located where your head was placed. This would overcome the forces holding your body together and literally tear you apart.

Einstein confirmed these calculations, which were finally proven in 1960 in an experiment at Harvard University. In the kind of star that Schwarzschild proposed, the forces would be immense. What is more, this star would be invisible to us—indeed it was called a "dark star" for many years because you would not see it against the blackness of space. This is because the downward force of the gravity well would be so immense that anything wanting to get out would need to travel faster than light to do so. As nothing could go that fast—not even light—nothing could escape— again, not even light. The star was invisible from the outside, or rather appeared as just a dark void.

The point at which a super heavy star crosses the threshold to become a dark star is known in honor of this young physicist as the Schwarzschild radius. In the mid 1960s physicist John Wheeler gave these dark stars a more poetic name—"black holes," which is what they would look like as seen from the outside.

Black holes cannot be seen through a telescope, but the forces they exert on nearby space-time can be a measure of their presence. Penrose explored the effects that would result from any such massive body with a gravity well that was incredibly deep. Here the distortion of time and space in their vicinity would be catastrophic. But more importantly, they also turned out to be an accidental time machine.

There is a point surrounding any black hole where it is impossible to go farther without becoming irretrievably snared in its grip. This is known as the "event horizon" because, quite literally, all events cease at that position, as light cannot escape it; time and its attendant sequence of happenings becomes frozen forever. If you were to send an unmanned probe towards a black hole it would seem to travel ever more slowly, then just stop moving as if it had run out of gas. In fact, time would have been slowed down and as the probe reached the event horizon, time would effectively have ceased so far as anyone watching the probe was concerned. The probe would still be moving, but so slowly and imperceptibly that it would appear, to all outside observers, like a statue marking the boundary point.

Curiously, as Penrose showed, the instruments on the probe would not reveal this. As it reached the event horizon and all hope of return was lost, nothing odd would seem to occur at all. There would not even be a boundary to cross. The probe would carry on towards the black hole and get sucked in by the gravity well, time passing normally as far as it was concerned. But it could no longer reverse course once past the event horizon, as there would not be enough energy in the universe to overcome the gravity well that was dragging it down. Nor could any message escape to the outside. The probe might take another half an hour to fly into the center of the black hole until the pull on its front was so much greater than the forces tugging on its rear that the craft would be torn to pieces.

Penrose noted that the probe would exist for a period of half an hour doing things and maybe recording data, which as far as the rest of the universe was concerned (or at least the one outside the confines of this super dense star), simply never happened.[29] It would be as if time had gone down two completely independent tracks in two different universes. Indeed, in many ways that would be exactly what was happening.

Penrose could show from his research that relativity and time dilation were being pushed to the ultimate extreme in these situations. The closer the probe got to the event horizon, the more time it took to go even very slightly farther. Every millimeter that it moved onward now took longer to travel as far as outside observers were concerned. At the event horizon it would still move, but in such a way that it would take quite literally forever for the next millimeter to be traversed. Light would have slowed down and stopped dead and along with it, time.

The work of Penrose would show that black holes were the most deadly things in the universe and would invariably create what is called a "singularity"—the point where everything—mass, light, time—is pulled into one infinitely small spot at the heart of this void. It is a bit like a super crushing device. His investigations had huge implications for the Big Bang theory describing the origin of the universe. Indeed, it emerged as quite possible that as matter is sucked into the singularity of a black hole, it might re-emerge in an opposing universe where everything then explodes out from that spot and expands across that universe, forming new stars and galaxies.

Was this how our reality began? Were black holes the engines that created new universes by acting like cosmic mincing machines? Did they suck in mass, energy, light, and time that came close enough to be gobbled up, then spew them out into another dimension to create new mass, energy, light, and time that kick-started a new cosmos? Black holes supported the Big Bang theory, but were harder to equate with Hoyle's rival hypothesis where time travel had no place. Certainly these weird natural phenomena do act like time machines.

Unfortunately, it is useless to just fly into a black hole and hope to time travel—you would never get out again and almost certainly would be crushed to death in its singularity. But Penrose reported a way in which you could play a dangerous game of tag in an attempt to time travel. If you flew close to the event horizon of a suitable black hole, but always stayed to the right side of its enticing lure, then you would find time in the universe speeding up more and more. Provided that you always left yourself enough fuel to fight the increasing forces needed to get away again, then you would be able to time travel in utterly spectacular fashion before you flew away.

For instance, if you flew in a probe towards the event horizon for just an hour and then headed back from the pull of the black hole, when you returned to your mother ship far more than an hour would have gone by. A century might easily have elapsed for those on board. As such you would be 100 years into the future and there would be no way home.

One problem would be that if you slightly miscalculated and went a little too close to the event horizon, then the time passing outside would increase massively. You might discover that a million years had elapsed or that the universe you knew had become extinct! Expeditions using these black hole time machines would be the most risky pursuit imaginable.

So black holes seemed rather hopeless as a practical way to break the time barrier—until 1965 when New Zealand physicist Roy Kerr found a way, provided the black hole was of a very special kind. He demonstrated that a black hole that rotated would form its singularity, not as a tiny point where everything was pulled in and crushed, but as a ring, surrounding the black hole like a donut. You could fly into the hole without touching the singularity and dip your toe in the regions of space where time travel is happening while still being able to escape.

This kind of black hole is a gateway to the future, a potential time machine that would let you choose how far you wished to travel, depending upon your precise flight path around the ring. Kerr had designed the first truly practical time machine that would let you move in minutes through months or even centuries. But it was still just a one-way passage. And it all depended upon finding such a black hole.

Even if you could find one, then purchasing a ticket on Kerr's time machine might well cost you your home, your family, perhaps your planet or even your universe. Unsurprisingly, there was no rush to go looking for a suitable black hole to put this idea to the test.

Nevertheless, science fiction writers were swift to see the wonderful stories that could be woven around the idea of black holes—even though these remained just a theory for several more years. Then evidence was found deep in space that suggested real, unseen black holes must be distorting things in their vicinity. None of them are remotely within reach of Earth, even traveling at the speed of light. But the proof for the existence of this extraordinary phenomenon has established itself in the popular consciousness, with many TV series and movies featuring the idea and even exploiting its time travel implications.

Even Hoyle got in on the act, although still declining to embrace the Big Bang concept that to many physicists went hand in hand with black holes. In his 1966 novel, *October the First is Too Late*, he envisages a series of parallel worlds coexisting in closely aligned spatial dimensions. These resemble the ruptures in space-time that would be created by the presence of a black hole.

The complex novel develops the math that would allow travelers in this type of universe to move a few miles through space (from one continent to another on his alternative version of Earth) and by doing so find themselves traversing time as well. Leave one country where it may be the twenty-first century, and arrive at another where people are living normally in the year 500 BC, when Ancient Greek philosophy represented the summit of intellectual development.[30]

The discovery of black holes had greatly upped the stakes as far as time traveling was concerned. Now there was a very real possibility that nature had built working time machines and they were out there in abundance waiting for us to go and find them. Nobody knew how close the nearest one might be, given that they were notoriously difficult to find.

But in the absence of an easily reached or suitable black hole, Frank Tipler, a physicist at Tulane University in New Orleans, seized the initiative. He was the first modern scientist to design a time machine that could be constructed in the laboratory rather than found deep in space, setting out precisely how this could be done thanks to black hole research. If we had the necessary technology then a Tipler Machine could be constructed tomorrow and it is widely accepted that it would almost certainly work.

Tipler's design for a working time machine was based on a smaller and artificially created black hole that could be constructed in the lab on Earth. One that you might then set into circular motion to create a ring-like singularity.[31]

But any time traveler located inside an artificial black hole would face massive difficulty and the black hole would consume huge amounts of energy. Its gravity well would have to be incredibly deep, with all manner of repercussions for anyone nearby. While there may be ways to contain such forces in a lab, it would certainly trap any time traveler who ventured inside and got too close to the singularity.

Tipler then tried a different tack. He discovered that if you were placed in a region of space and an artificial black hole constructed around you, it could be possible to time travel, but you would never be able to convey knowledge of that success to anyone. This is because the high gravity well would imprison you and trap any message that you might seek to send out. Nor would you be able to summon up enough velocity to escape. And lastly, the forces would very likely crush you to death.

However, Kerr's calculations showed that a rotating black hole should create regions of space in its immediate vicinity where time was distorted enough to allow some time travel without the same devastation. These would be like eddies in water around a rock in a stream. If you circled the edge of this artificial black hole inside a machine that twisted around its circumference, then the crew would pass through these temporal eddies and travel through time. The more times you twisted around the black hole, the further through time you should be able to travel.

Tipler's device is curiously similar to Drouet's time rocket. But the difference is that Tipler's is tailored to match what we know about the physics of black holes. The artificial black hole would probably be too dangerous to manufacture in a laboratory. A safer option would be to create it in space, far from any planet. And the time machine that circled it would need to be a form of flying craft akin to a space shuttle. His time machine would be a spacecraft.

The optimum form of Tipler's time machine would be a long cylinder or vortex—and by long he certainly meant it. It would need to be over 60 miles long, and be made of a very dense material. Such a device would induce huge stresses, and how to create such a dense mass remains the major problem. This blueprint for a working time machine soon hit a brick wall as the practical difficulties mounted—especially as massive amounts of energy would be needed to build it—possibly greater than the sum total yet available on Earth.[32]

Tipler's time machine was a worthy effort but is not likely for the foreseeable future, and the first experiments would be among the most dangerous that mankind has ever contemplated.

1973
BACK FROM THE FUTURE

History is riddled with stories about madcap inventors coming up with technology far ahead of its time. But nothing is likely to be any stranger than the claim that the Vatican came into possession of a time machine following research by some of the physicists who built the atom bomb.

Two of the scientists supposedly behind this "chronovisor," which means literally "time-viewer," were the celebrated nuclear physicist Enrico Fermi and the space rocket pioneer Werner von Braun. But there is no direct testimony from their memoirs or records of any such work done by these men. It was Pellegrino Ernetti who revealed their alleged key role in this affair, saying that he coordinated their theories and then built a device for them. But Ernetti was himself a curiosity, as is his entire tale.

Ernetti had a degree in quantum physics, but was also a Benedictine monk based in Venice, where he authored numerous scholarly works. During the 1950s and 1960s Ernetti planned the construction of a chronovisor with full support from Fermi and others, he says. Towards the end of his life, he told all in interviews, but he was otherwise very circumspect in describing the apparatus or in allowing others to see it. He stood by his story that the research was real.[33]

The chronovisor did not transport anyone into the past, but captured images from another time period to be viewed on a screen. It supposedly worked using microwave energy that manipulated the frequency of sound and light waves in order to alter how far back in time images were coming from. Only images from the past were possible since the future was unformed and possessed no residual energy that could be detected, amplified, and brought into the visual spectrum by this device.

Using the chronovisor, Ernetti witnessed numerous historical events unfold, such as scenes from the forum in ancient Rome. But the scholar published only one reputed "screen grab," a backlit image of a bearded face staring towards heaven. When this supposed picture of Christ during his crucifixion hit the Italian media in 1972, it was greeted with furor and widespread doubt. Most thought it too similar to a wooden crucifix located in Perugia, and the monk was accused of fraud. Those who knew him attested to his integrity and argued that his reputation was such that this kind of deception would have been foolhardy.

In 2001, an American publisher received an anonymous note alleging that Ernetti had made a deathbed confession. The photo of Christ was a fake and the chronovisor did not work—but the device was built and the theory behind it was sound. To his death, the monk believed it would work.[34]

Following the release of this unsigned note, François Brune, a French priest and longtime friend of Ernetti, added yet another twist, stating his absolute belief in the monk's sincerity. Moreover, he alleged that the Vatican had been deeply concerned by Ernetti's device and had ordered the time traveler into silence out of fear that his viewer could be used for ungodly acts. So worried were the papal hierarchy by the achievements of this physicist that they ordered Ernetti to dismantle the viewer and scatter its pieces to different locations around the world to preclude its reconstruction after his death.[35]

Could a group of physicists have built, after World War II, a viewer that could see into the past? Remember that the Eglin Air Base test was of a device that did something similar in the 1950s. And recall that our telescopes also look into the past; images coming from 1,000 light years away reveal the stars as they were 1,000 years ago, because that is how long these rays of light have been in transit across the universe. So in some sense the Hubble telescope in Earth orbit is very much like a chronovisor. We can tune it to observe the nearest star (Proxima Centauri—just over four light years away) and then look at the Andromeda galaxy (an amazingly distant 2.2 million light years). In doing so, we will see images coming from vastly different periods in the past one after the other.

Theoretically, we could tune this telescope into any past time, be it eight minutes ago—the time taken for light from our sun to reach us—or a period when no human beings even existed on Earth tens of millions of years ago. The only difference is that we cannot use a telescope to see what occurred here on Earth, we can only pick up the light just arriving after a very long journey through space. But the principle is much the same as Ernetti described and does not stretch the bounds of scientific credibility.

So how might we try to create a device that records light from a past event on Earth? One way would be to find light that somehow gets trapped in a loop or has gone the "long way around" on its journey to reach us, just the opposite of the theorized shortcuts allowed by a wormhole or an Einstein-Rosen bridge. Instead of a quick nip through another dimension to reach its destination ahead of time, it is just as likely that some paths through other dimensions might increase the time required, so that something expected to arrive yesterday might not arrive until the day after tomorrow. We could wait for it, tune in, and see the past.

It is perfectly feasible for a wormhole to exist that would increase, not reduce, the distance between two points in space. We are naturally intrigued by shortcuts through space because of their ability to let us beat light and thereby time travel. But statistically there is no reason why many wormholes would not take the long way around, thereby slowing down the passage of light rays from any past event. If so, then can we tune into light that has been delayed on route to let it be recorded and replayed at some point, perhaps centuries after the event that it conveys? Light from all great moments in time might be out there, circling slowly down some wormhole detour heading home.

Another way to design a time viewer to have access to images from the past would involve finding some signal that can travel faster than light. Anything that beats light traveling through the vacuum of space must also travel backwards in time. If anything does exist that can surpass light speed and we find a way to control its use, we could then reach back into the past and lock on to light rays from that period, perhaps returning them to the present. This could allow us to develop a viewer that sees into the past.

Einstein believed that nothing could travel faster than light, so this suggestion had always appeared doomed. But he was mostly concerned about any object traveling at light speed, not faster than it, because this introduces apparently nonsensical results into the relativity equations. Material objects would possess infinite mass, for instance, which is self-evidently absurd. Of course, that word "material" may be the key, for non-material things (such as energy waves or particles of light) clearly can travel at light speed. Consequently these mathematical issues can be overcome in certain circumstances. This loophole has increasingly been seen by modern physics as a possibly indirect way to allow light speeds, or maybe faster-than-light speeds, and thus to time travel.

In 1801 the British physicist Thomas Young had first tried to discover whether light rays were waves of energy or streams of particles. It proved very hard to decide. When he shone a beam of light through a tiny slit on one sheet, he found that it left a patch of brightness on the sheet behind it. This could be adequately explained as a flow of particles (the photons) that were passing through the slit and flooding the sheet with light. But then Young used two tiny slits side by side, and the problems began. You might now expect two patches of light, one from each slit, but instead there was a line of light and dark bands that can only happen if light is being transmitted through the slits in waves. Where the waves interact, they either reinforce or cancel each other out according to whether their peaks and troughs combine. So light seems to behave both like particles *and* like waves.

During the 1970s it became possible to do a very sophisticated upgrade on Young's double-slit experiment, and to the horror of many physicists things became even more confusing. They discovered that you could actually affect the outcome of the experiment by changing the way you take the measurements. If you let the photons fall one at a time onto a projection screen, the interference pattern always appears when you use two slits. But if instead you place a detector to record the photons as they pass through each slit, then something incomprehensible occurs. You now get two patches of light, one behind each slit. There is no interference pattern. You have altered reality by making a decision on to how to make the observation. Let the experiment run its course and one outcome emerges, but force the photons to be counted one by one as they pass through the slits and a quite different outcome appears.

As if this wasn't bad enough, then came the weird version of the double slit experiment set up by the University of Munich, which was replicated by scientists at the University of Maryland. Their experiment, called the "delayed choice experiment," used detectors, one behind each slit, ready to count the photons that went through the respective holes. However, you can do this experiment in three ways. First, if the detectors are switched off, the photons are not counted passing through the slits and, as expected, an interference pattern forms. Second, if the detectors are activated, then there is no interference pattern, just a patch of light produced by each slit. And third, the detectors are turned off, the photons pass through the slits, and only then, after the experiment has run its course, are the detectors switched on. What result would you now expect this experiment to produce?

Amazingly, if you choose to switch on the detectors *after* the experiment, there are two patches of light. If you do not switch them on, there is an interference pattern. This is what happens even if you make this choice long after the photons have passed through. It seems almost ridiculous, as if you are changing the past depending upon the choice made well after the experiment has ended. How can any decision that you take after the photons have clearly already passed through the slits possibly interfere with things that have long since happened? The only apparent conclusion is that time travel is taking place.[36]

It is as if your choice sends a signal back through time from the moment that you make this decision, alerting the photons as they pass through the slits to let them then decide what they must do. Or, to be more exact, while our method of observation dictates the outcome of the experiment it does so independently of time. We can choose which way to view the result before or after we have conducted the test. The result will still be the same.

On the face of it, this new twist on the double slit experiment shows that photons of light can somehow time travel from the future into the past. The dreaded idea of changing the past by actions made in the future was a massive culture shock, even to modern physicists used to quantum weirdness.

The notion of particles operating in a timeless state opened up a new debate. Several scientists proposed that there might be a whole range of particles that can only travel faster than light and never slow down below that limit, just as we cannot normally propel matter fast enough to reach to that limit. Such faster-than-light particles are now known as tachyons. The hunt for tachyons became one of the hot new items on the time traveler's agenda during the 1970s.

Although the existence of faster-than-light particles is highly controversial, any message sent by a beam of tachyons would travel into the past, where, of course, it would seem to come from the future. Tachyons provide a ready-made chronovisor-like time machine. Unhappily, calculations by Princeton astrophysicist Richard Gott showed that these particles self-destruct as their velocity exceeds light speed by increasingly greater margins. For this reason, nearly all tachyons that exist, if indeed they do exist, would move at just above light speed. While they could indeed send data back through time, it would probably only be over subatomic distances and through very short periods of time.[37] But this research excited Gott enough that he became a would-be time traveler, and he is today one of the leading figures in the physics community seeking to find a way to build a time machine. His paper on tachyons in 1974 led many scientists to enter the race.

Gott's paper in the *Astrophysical Journal* had a not exactly catchy title—"A time symmetric, matter and antimatter tachyon cosmology"—but it legitimized the idea of things that might travel faster than light and discussed how they might thereby be used to construct a practical form of time viewer.[38]

Gott's work inspired physicist Gregory Benford, a particle researcher in California who writes science fiction. In 1980 his award-winning novel, *Timescape*, even included Gott's paper in the plot. Benford realized that the best way to use tachyons for time travel was not to try to transport humans into the past, but to send information as an energy beam backwards in time by making use of tachyon-driven carrier waves.[39]

By the early 1970s dozens of different kinds of particles were known to exist in subatomic space. Some were long lasting and stable, but many were very tiny, short-lived, and prone to self-destruction—just as Gott found tachyons would do when pushed far beyond light speed. The particles discovered in decades of experimentation all obey the accepted rules of physics and can never actually achieve the speed of light, let alone surpass it. Only massless particles that transfer information—such as the photon—travel at light speed. Nothing with mass possibly could do so, even theoretically.

However, Einstein's equations, when you consider potential faster than light motion, have some curious negative numbers in them that imply there might be a whole school of particles with behavior mirroring those that form normal matter. Most physicists think that this is probably just a mathematical abstraction. But if these hypothetical solutions do mean something, then there could be particles, such as tachyons, that never slow down, even to light speed.

The celebrated physicist Gerald Feinberg began the search for tachyons in the 1970s by scrutinizing what happened during collisions between particles inside linear accelerators. During engineered collisions between subatomic particles inside these accelerators (or supercolliders, as they are sometimes called) new particles are occasionally produced. A few of these have remained mysterious. One or two may have been tachyons traveling faster than light. But there are usually other possible solutions, and no tachyons have been proven to exist as yet.

An experiment conducted in Australia in 1973 seemed to produce the first plausible evidence of faster-than-light particles, although many scientists would disagree with this result. The experiment used photographic plates to record high-energy particles bombarding the Earth in cosmic ray showers. But something left traces on the equipment that seemed to come from unidentified particles that arrived there before the cosmic ray shower had even begun.

One way to interpret this curious result was to suggest that the plates were disturbed by particles traveling faster than light and so going backwards in time. This caused them to arrive on Earth before the rest of the shower set off, since it was restricted to traveling at light speed.[40]

Proving that tachyons exist may hinge on the reality of antimatter, which has mirror properties to normal matter. Indeed, it is now widely accepted that the best way to interpret a positron—which is the antimatter version of the electron with identical mass but opposite charge—is to view it as an electron that moves *backwards* through time. The Caltech (California Institute of Technology) physicist Richard Feynman, who is regarded as one of the great minds of the twentieth century, worked on the Manhattan Project to build the atom bomb and investigated the Challenger shuttle disaster in 1986, but was awarded the Nobel Prize for his discovery of this concept of time reversing subatomic particles.[41]

While tachyons remain elusive, there is a growing belief among physicists that this could be one of the easiest ways to perfect time travel. Perhaps, as Gregory Benford's novel suggests, one of the first successful time trips may well involve the sending of energy beams through time that conveys a message and might interfere with radio transmissions, TV signals, or computer data. Science fiction often paves the way for real developments in time travel physics. If we heed the ideas in Benford's novel, then perhaps we should start looking for this kind of interference in our equipment. If we find anything amiss, then we would need to assess the cause. That niggling problem that disrupts our communication tomorrow or the day after tomorrow might turn out to be a glitch deliberately sent our way from the future.

THE TIME MIRROR

1983

By 1983 the brilliant scientist Stephen Hawking was getting rather worried about time travel. Ongoing discoveries had revealed how nature was able to time travel virtually at will, and results from experiment after experiment caused him great discomfort. Although he knew it should not be possible, several of his colleagues were busily trying to create a device that might carry people through time.

Since his days at Cambridge learning physics under Hoyle and Penrose, Hawking had been asking some very perplexing questions. If black holes can be manipulated or manufactured in the lab to act as a time machine, then surely someone, somewhere in the universe must have developed the capability by now? Or, at least, one day, he argued, someone will. If so—then where are they? Time travelers from some alien civilization or someone coming back from our own future when this method has been perfected should be zipping around the cosmos via a temporal freeway network of artificial black holes. Surely some would visit our time, or have journeyed through recorded history. So where were these visitors, Hawking wanted to know?

Hawking was working on his magnum opus, *A Brief History of Time*, which would become the most widely purchased physics book ever produced for a mass audience, spending years on the best-seller lists.[42] He had already proved, contrary to expectation, that black holes would emit a specific kind of radiation rather than, as was previously supposed, allow no data of any kind to escape their clutches. The reason? Complex subatomic reactions that occurred near the event horizon would produce one set of particles that would be sucked towards the singularity, while a mirror set would escape outward. This meant that black holes would not be completely invisible. The ejected radiation (named "Hawking radiation" in his honor) should lead anyone hunting for them directly to a candidate black hole.

Much later, in July 2004, he even told a conference held in Dublin that the very concept of a fixed event horizon was misleading. In the 1980s he had believed that beyond the event horizon no data from any object that enters a black hole could ever escape, but he now accepted that this was a mistake. The word "never" when applied to black holes really should mean "a very long time." Eventually a black hole would reveal the secrets of anything inside. Time would not stop in its vicinity; it would just slow to an extreme crawl.

Hawking took a few more years to ponder the meaning behind the apparent dearth of time travelers perhaps using black holes to travel to appear in our midst. He would return later to the puzzle and add a new twist to the time travel story. Meanwhile another student at Cambridge had come up with a clever experiment of his own to test the time travel concept.

John Lucas was then studying mathematics and in late-night drinking sessions with other students debated the problems created by the laws of space-time and how their equations seemed to allow for time travel. Lucas hit on the bright idea of opening up an investment account with a modest sum of money that was easily affordable, even to a student. The plan was to leave it in the bank for as long as it took for time travel to be perfected—even if that was decades or centuries. By then the interest should have made the sum sufficiently large to enable the purchase of a time machine that would be sent back to a prearranged spot in the 1980s. If time travel was ever going to become real,

Lucas argued, the time machine should therefore already be in that spot waiting for him—having been purchased in the future with his own money, perhaps by his great, great, great-grandchild!

Needless to say, the scheme did not work. Lucas admits that no money was ever actually invested because his thought experiment concluded that the time machine should arrive immediately even when nothing more than a firm decision to open the account was taken. Given that this decision was taken and the time machine failed to material-ize, it was pointless going through with the investment.

The experiment raised numerous questions, ranging from whether you could buy a time machine even if time travel becomes possible in the future, to matters of physics that may well deny such a machine returning through time to set up a temporal paradox. You might also argue that since the money was not invested, no firm decision was really taken, and as a result the experiment was self-defeating from the outset.

Nevertheless, this experiment indicated how time travel had begun to capture the imagination of academics. Lucas, meanwhile, gave up trying and turned his frustrations into a comic science fiction novel called *Faster than Light*.[43]

Frank Tipler had discovered a fundamental problem with experiments such as the one by Lucas when he had de-signed a time machine based on an artificially created black hole. Such a device would only allow travel back to a period *after* its own construction. In other words, if you were to build the prototype in the year 2020, then a decade later it could be used to return from 2030 to the year 2021, but it could never go back to 2019 or any time earlier than that. This may well explain why future time travelers are not all around us. *Our present time is unreachable because it predates the moment when the first time machine is activated.*

If this conclusion is correct, the success or failure of such an experiment as Lucas's cannot be judged until the first working time machine is actually turned on. It also means that nobody will ever be able to invent a time machine in secret, because from the first second that it starts to work, future versions of itself would start appearing all over the planet like daisies in the spring!

While Tipler was hitting the twin brick walls of practical impossibility and the restrictive laws of time travel, Princeton's Richard Gott had developed another idea for a time machine. He dubbed his very clever proposal, which was based on the notion of time delays, the "time mirror."

Anyone who has ever used a telephone on a long distance call with the messages relayed via satellites, or, worse still, a television satellite link used to interview people on another continent, will know the trouble caused by the dreaded time delay. This is induced into such communications because electromagnetic waves take a finite time to travel. The delay of just one second in relays to and from Earth-orbiting satellites produces very real communication difficulties, breaking up the natural flow of conversation. We see this effect in TV news broadcasts every day.

Of course, there is a time delay in absolutely everything that we experience, which is caused by the fact that light takes time to travel even small distances. When we look in a mirror, we see ourselves not as we are now but as we were an infinitely minute fraction of a second before, because the light has had to travel to the mirror, reflect back off it, and return to our eyes. If we see something ten miles away across a city landscape, we are seeing that spot even further back into the past, although again only measured in fractions of a second and we never notice the effect. But all acts of vision involve time travel to some degree.

Gott used this fact to design his time mirror. Imagine that we could send a powerful light collector to a black hole one hundred light years from Earth. Gott would have us use this light collector to power the light rays curving in an arc around the black hole back to Earth. Why would this be of any use for time travel, you might ask? Put simply, because it would open a window on yesterday.

It seems like a lot of effort for little reward, because those rays would travel at the speed of light and take 200 years for a round trip to the black hole and back to Earth. If launched tomorrow, such a device could not be made operational until the year 2205 given the journey times required. However, once it was up and running, the people of the twenty-third century using this mirror would be getting images of our planet that left here in the year 2005 and had undertaken this massive round-trip. The time mirror would be showing them their past and revealing in exact detail what our present world was like.

If many black holes exist in the universe, as physicists think is increasingly likely, then we can predetermine how far back into the past we choose to look by positioning the appropriate time mirror at the correct distance from Earth. A black hole fifty light years away would show the Earth of one hundred years ago; one that is 1,000 light years away would delve two millennia into the past.

Constructing Gott's time mirror would be an extremely difficult technical achievement. At such vast distances, light rays will be very feeble. It will be exceptionally difficult to separate them from the overwhelming glare of a nearby sun or boost them enough to ever be detected on their return to Earth. But this kind of time travel breaks no rules of physics—upsetting neither the speed of light nor the old bugbear of cause and effect.[44]

Yet even this version of a time machine suffers from the limitation factor foreseen by Tipler. You can never see the past prior to the dispatch of the time mirror. If we built one today and transported it to a black hole one hundred light years away, it would take at least one hundred years to get it there. It also needs another one hundred years for its first signals to be received back on Earth. So in 200 years we could start viewing the past, but the first images would, through the laws of physics, have to date from *after* the moment that we first sent the time mirror towards its destination. It could never be any earlier than that.

The 1980s proved to be a frustrating period in the race to build a time machine. Again and again Tipler's limiting factor brought efforts to a halt. Time travel might have a built-in restriction that could forever prevent us from sending a camera back to take pictures of living dinosaurs, or the myriad other uses scientists could find for any kind of time machine.

Tipler seemed to have explained why the universe was not awash with time travelers, as Hawking had perceptively thought it ought to be, and at the same time he had demolished all of those time travel novels about going back to save President Kennedy from assassination. Time travel into the past might be possible, but not travel into our past— only the past of the days of existence following the construction of the first time machine. And at the moment this past remains in our future.

1984
ENTANGLED TIME

In the 1980s the construction of a time machine was at an impasse with snags arising at almost every turn. Success seemed a long way off, despite all the theories and experimental results suggesting that it must be possible.

Black holes and time mirrors needed technology well beyond our current abilities. Wormholes might exist, but we had no way of traveling fast enough to go and find them. The 1973 discovery of faster-than-light particles could provide the breakthrough, but Gott had shown that these tachyons had major limitations for sending messages from the future into the past. But all was not doom and gloom.

There was another possible way forward. Time travel

appears to happen constantly in nature on a microscopic level. However, great minds like Einstein had never been happy with this sort of time travel resulting from quantum theory, exacerbated by the double-slit experiments that seemed to show that the past could be altered by events in the future. This worrying scenario had led Einstein in the 1930s to design a thought experiment, along with colleagues Nathan Rosen and Boris Podolsky, which aimed to expose the sheer implausibility of such microscopic time travel methods.

Einstein found it absurd that for quantum physics to work, subatomic particles would need to travel faster than light and move into the past. Quantum theory says that particles from a common origin would be linked, or entangled in the language of physics. Change a parameter of one particle and its sister will instantly change as if it knows exactly what is happening, regardless of where that other particle is now located. They would appear to communicate instantly across infinite distances using no known method to transfer this data and forcing changes after any unpredictable event to ensure that these entangled twins acted in unison.

It is easy to see why Einstein was certain that this interpretation of quantum theory was nonsense and why he hoped that his thought experiment—if it ever could be carried out—would prove this. Otherwise the universe would be awash with linked particles, and time and space would be an illusion. Faster-than-light communication back through time would be commonplace. It is little wonder that Einstein expected to one day be proven right.

The development of particle accelerators in the 1950s offered that opportunity. They could smash particles into one another, creating entangled pairs for use in lab tests. Once this technology was perfected, Einstein's bold experiment could be carried out for real.

By 1974 the accelerator built in California was sufficiently calibrated to allow for the fine measurements required to prove that Einstein was wrong. The entangled particles behaved precisely as quantum physics predicted they should. Changing the spin of one particle altered the behavior of the other, involving communication of the necessary data without any known energy being involved. Einstein had often used a derisory term for these seemingly ridiculous effects—"spooky action at a distance." He would have been stunned to see it really happen.[45]

But the results did not reveal whether faster-than-light communication was involved—only that some weird contact happened. Faster-than-light communication had really worried Einstein because it contradicted relativity.

Then between 1982 and 1984 Paris physicist Alain Aspect, working with a team that included Jean Dalibard and Gerard Roger, ran some sophisticated new versions of these experiments and published their extraordinary results. The trio had figured out an ingenious way around the problem faced by previous experimenters of limited distance within the lab.

The team could easily create twin particles (in this case from calcium atoms) and send them in opposite directions down a sealed tube. But that tube could not be made many miles long, let alone test whether instant communication could occur through light years across the cosmos. What they could do was vastly increase the response time that could record how swiftly any changes made by one particle were mirrored by the other particle.

The accuracy of this measurement had to be greater than the time it took any signal to travel at light speed along the length of the tube. It would then be possible to detect whether faster-than-light messages were being sent between the two particles. The Paris equipment was able to measure down to one ten millionth of a second, which was sufficient, given the length of tube, to test whether communication between the entangled particles was occurring at faster-than-light speeds.

In the Aspect experiment, one photon of light from the calcium atom was fired towards one end of the tube and its entangled twin was sent towards the other end. At each end receptors were calibrated to precisely record when each photon arrived. But these photons also passed through pairs of polarizers (much like those used in sunglasses). Dependent upon their patterns, these polarizers would alter the behavior of the photon before it was detected and cause it to be aligned in one of several possible ways when striking the detector. The experiment would monitor whether entangled pairs changed in sympathy and how fast they adapted.

If quantum theory is wrong, as Einstein believed, then the twin particles would possess different properties upon arrival because they had passed through two different polarizers that had randomly changed their orientation. If quantum theory was correct, the twin particles would be entangled and act as one. Changes induced by passing through the polarizer would be transmitted to the twin photon, causing it to react in sympathy. The detectors would reveal how fast the switchover occurred and whether any communication between the photons was occurring faster than light.

Test after test produced the same outcome. The entangled particles were spookily linked and any communication causing them to act this way not only exceeded light speed, it appeared instantaneous.

Although experiments like Aspect's have now proven quantum theory beyond a reasonable doubt, Einstein was by no means alone in denying this version of the nature of reality. Erwin Schrödinger was also highly skeptical. His thought experiment imagined a cat inside a sealed box and a poison pellet that would kill it. These pellets would release a toxin if, and only if, a subatomic decay triggered it. If it did not, the cat lived.

There is, of course, a 50/50 chance of either event occurring. To find out which of the two options result, we would presumably open the box and see if the pellet has fired and if the cat is alive or dead. It is hard to believe that such common sense is tossed out by quantum physics. But it is.

That's because quantum physics predicts that the pellet fires and the cat dies *and at the same time* the pellet does not fire and it survives. You have to envisage a split personality live/dead cat existing in some weird limbo rather than just one outcome taking place as any sensible interpretation decrees. These two possibilities are called superpositions and exist until something causes reality to manifest as a live cat or a dead cat. This manifestation is known as "collapsing the wave function"—in other words, making the choice between the various possible outcomes.

Like Einstein, Schrödinger argued that the crazy outcome of his experiment proved that quantum physics was in error. In the real world, of course, things would be far more complicated than the simple choice of the live/dead cat. Every event that happens involves countless possible outcomes, and trillions of particles interact in a myriad of different ways even during the most straightforward incident. Yet quantum mechanics argues that all these multiple states must form superpositions and all remain real until the wave function collapses into the reality that we experience.[46]

We even have direct visual evidence that this occurs. Don Eigler and a team at the IBM lab in San Jose used a scanning tunneling microscope to capture an image of individual atoms. However, in the process they saw what they termed a "quantum mirage" that revealed the presence of the electrons inside the atom. But these electrons were not where the real atom was located. They were in one of the superposition states where the particles might have existed had another potential outcome been fulfilled. This state had not actualized, but this phantom state was real enough to have physical properties that the scientists hope to use in future applications to massively boost computer processing speeds and memories to levels that are not remotely feasible today.[47]

Physicist Hari Manoharan refers to the quantum mirage as "half real, half ghost." But all joking aside, there is little doubt that the superpositions revealed by quantum mechanics have a real existence. That poor cat might well be both alive and dead at the same time.

How we evaluate Schrödinger's experiment is really a fundamental question facing physics today. There are several options—each with extraordinary implications. And this field of research arises directly from quantum entanglement for which Einstein may be said to be its reluctant discoverer.

American physicist Hugh Everett proposed a completely different answer to what these states might be and it unexpectedly turned out to have major implications for time travel. He argued that these superpositions are all real and each one creates its own new universe. Whenever a sub-atomic event with multiple possible outcomes occurs, all of these outcomes become real and form a series of parallel universes that have a separate existence and development. Our consciousness inhabits just one of them—the reality that we experience. But every option emerges into its own version of the cosmos where alternative versions of "you" experience that particular outcome as real.[48]

This theory avoids the need for anyone or anything to collapse the wave function (because they all collapse) and explains why we might see mirages of other states—those closely aligned to our own may overlap. For that reason it appeals to some scientists, such as Hawking. However, its consequence is that trillions of new universes are being created every moment as a result of all the events taking place across the universe. Such a theory seems as ridiculous as the outcome where the cat is dead and alive at the same time, which is why plenty of physicists reject it.

Yet the "many worlds" theory, as it is known, is hugely significant for time travel. If correct, there is no longer just one universe to explore, but trillions of them. Some of these would very likely have developed conditions that favor time travel. Indeed, relativity argues that circumstances do theoretically exist that would facilitate time travel. So, if there are many different universes coexisting, time travel will be a common occurrence in many of them. We may just need to access one of the universes where this is true.[49]

Roger Penrose, of black hole fame, has proposed a further theory about how the wave function might collapse. He argues that superpositions are all perfectly real but only persist at the quantum scale. With larger objects, gravity forces a single reality to collapse out almost immediately. The deeper the gravity well the large object possesses, the faster this dissolution of virtual states occurs. But subatomic particles are so tiny that their gravity has negligible effect and their superpositions last long enough to seem to be permanent. However, when objects are greater than a certain size, the increased gravity will always trigger the superpositions to dissolve too swiftly for us to see them. This means anything remotely as large as a cat (indeed even objects as big as atoms) will have superpositions that exist for a tiny, tiny period of time and we never detect them. They rapidly collapse into a single reality state—the universe that we inhabit.

But even if Penrose is right, then everything in the universe still exists in a range of different states just for a fraction of a second. There would be billions of superposition states side by side in alternative versions of reality literally filling the cosmos with phantoms—even things the size of people and planets. We must all have split personalities that inhabit multiple universes.

Quantum physics has exposed a view of reality that is truly weird, and where time travel may be the least of all the strange things going on. To time travel we may not need to reach for the stars, but only look closer to home. Perhaps even inside ourselves.[50]

1988
WORMHOLES

Deborah Jin at the University of Colorado had informed a gaggle of bemused reporters that she had just done something wonderful in her lab—create a new state of matter known as a Fermionic Condensate. Needless to say, few understood the importance of her words.

This gaseous substance is named after the Italian physicist Enrico Fermi, who worked on the atom bomb during World War II and is reputed to have assisted in the creation of the controversial Vatican chronovisor. Her news probably seemed of minor interest to the few media sources that covered the story. But it was a key moment for the design of a time machine.

The condensate created in Colorado is not really a solid, liquid, or gas, but a dense medium that seriously impedes the transfer of any data. When created at very cold temperatures, it smears out to form a super atom with unusual properties. It shares features with the Bose-Einstein condensate that produced stunning results when it was first created (see the preface). Other examples, possessing a range of chemical and physical properties, are likely to be produced in the lab in the near future. Although these condensates are not being produced to facilitate time travel, they do affect the flow of light waves and sooner or later one of this new array of weird substances may have just the right combination of properties to provide the fuel source for easy access to time travel.

The Fermionic Condensate was developed to provide a suitable storage medium for a computer system. Massive increases in processing power can be generated if light waves can be made to replace microchips to hold data, but for that to happen light needs to be readily controlled, and the condensate may be the answer. Unfortunately, we are still some way from designing room-temperature models, and computers working at bitter cold levels seem equally impractical. Nonetheless, slowing light is the way forward. But it has some strange side effects.

Slow light will make time pass more slowly. If we can trap the light of past events inside a condensate, like lightning in a bottle, history may be relived at the press of a button.

Condensates are exotic matter, allowing some limited control over light and time. But it is another kind of exotic matter that may prove critical to any plan to build a time machine—especially a plan hoping to bring about travel into the past. Or so says mathematical physicist Paul Davies.

Davies has developed a serious reputation for his bold and farsighted thinking about the increasing likelihood that we will find a way to time travel. He thinks that wormholes might be the key.

When Einstein and Rosen first introduced the possibility for these natural bridges through other dimensions, the concept was just an expedient to overcome the damning time travel implications of relativity. As such, they remained of minor interest for a very long time, until the race to build a time machine began to warm up in the 1970s. By then wormholes were seen to be the best hope to allow human travel into the past, whereas almost every other method under discussion allowed only travel into the future, or, like the chronovisor, possibly might bring just images or data from the past.

Wormholes are certainly consistent with relativity and were predicted by it, but nobody has ever observed one. Unlike black holes, where physical evidence for their reality has been found, wormholes remain the most sought after natural phenomena in modern science.

One reason for the importance of wormholes is that

they may be used as nature's time machines. Wormholes and the time travel physics that they bring excite physicists more by revealing the nature of space and time at the very edge of our knowledge and may be the best way to comprehend how the forces of the universe mesh together. To finally discover the TOE (Theory of Everything) we might need to understand how wormholes work.

The natural tendency of a wormhole is to collapse under the huge gravity forces that are involved. Why? Because the forces that operate inside the wormhole once its tunnel is open are enormously stressful and the math reveals that almost every imaginable form of wormhole would be inherently unstable. Indeed, the mouth of the wormhole may stay open for a very short length of time. This would be disastrous for anyone hoping to time travel by passing through it—the collapse is likely to close off the exit point before you can get there. It also means that wormholes may be closed off most of the time, making their discovery rather like finding a needle in a haystack by hoping that the sun will glint off the needle for a brief moment at some point during the day. How do you know where or when to look?

Davies calculated that what was needed was a cosmic pitchfork to pry open the mouth of the wormhole permanently. Exotic matter may be the one thing that can do that because it can effectively produce the antigravity needed to counteract the massive gravity forces that would tend to slam the wormhole shut.

Fermionic condensates do not produce antigravity effects, so they are not the right kind of exotic matter for wormhole control. But the fact that they can now be manufactured, along with the Bose-Einstein condensate, shows that the search for a form of exotic matter that might bring antigravity and could stabilize a wormhole is not a fantasy.

Davies believes that this might be possible for us in the future, or even possible today for an advanced alien civilization. If such an advanced race can keep a wormhole stable, Davies has calculated just how they might turn it into a time machine.

First stabilize the wormhole. Then tow one mouth of the wormhole through deep space so that it comes very close to a dense object such as a neutron star. The star's enormous gravity well would then produce massive distortions in space-time and eventually open up a time tunnel. The gravity forces involved would cause a discrepancy between the time experienced by the mouth of the wormhole next to the star and the exit mouth left elsewhere far away from the gravity well.

After generating a time difference between the two ends of the wormhole (which could take you backward for any number of days, weeks, months, or years depending upon how long you left it next to the neutron star) the wormhole mouth can be towed away. Parked wherever you needed to use it—perhaps near the moon to make it easily accessible from Earth—you would now have a reusable time machine with a set period of time transfer involved.[51]

Such a device obviously involves huge engineering difficulties—not to mention small matters such as discovering wormholes, creating the right kind of exotic matter that furthers antigravity, or proving that any of these things exist in the first place. But a machine that could convey humans into the past is theoretically possible.

Cosmologist Carl Sagan used many of these ideas in his 1985 novel *Contact*, which was the basis of a hit movie a decade later. He imagined a message from a distant star system being received on Earth and containing the blueprints for a spaceship that allowed one person to travel across the galaxy to its origin in the star system we call Vega. However, when the scientist selected as the first intergalactic voyager returns to Earth, she has great difficulty persuading anybody that she has really traveled many light years, because they insist that she has never even left the launchpad.[52]

There is good reason for their apparent doubt and it results from just how Sagan designed this extraterrestrial craft. His alien spaceship works by creating a temporary rip in the fabric of space to forge an artificial wormhole through which it can travel, but it turns out to have a major side effect. Time and space are inextricably linked, so this ship is not just a space traveling vehicle but also a time machine. As a result, the journey across trillions of miles was over before it had begun from the perspective of anyone on Earth. Indeed, the bridge through space that was created by the ship caused the astronaut to travel back in time during the journey, allowing her to arrive home at a point slightly before she had actually departed! She experienced the voyage to another world occurring in normal time, but to observers on Earth she seemed to have never left. Time travel might well really have such bizarre manifestations.

Sagan was careful to check his science with specialists who research the nature of time, one of whom was physicist Kip Thorne at Caltech. This fictional alien wormhole machine was based on Thorne's suggestions. But so intrigued was Thorne by collaborating with Sagan on his science fiction story that he used it to try to design a real device that could operate in just the same way.

While Thorne knew that we could not build such a craft in 1988, when he published what was in effect the proper blueprint for his time machine, he showed that it was completely possible. From that day forward constructing a time machine could never be said to be beyond the realms of modern science. It was only a matter of time.[53]

Imagine an intelligent race of beings on Alpha Centauri who possess a super telescope that lets them observe events on Earth. They will be watching things that occurred here four years earlier, because that is how long it takes for light to travel to their star system. But if they find a wormhole linking these two points and beam "live" transmissions of themselves watching Earth, then it may take only a day for that light to reach us. We will then see one-day-old pictures sent from Alpha Centauri revealing the locals viewing "live" images of our planet four years back into the past. We have been gifted a machine to view our own past as it unfolds.

If we now fly through the wormhole towards Alpha Centauri, then time (as conveyed by light taking the "slow" route) will continue at its usual rate. So when we arrive on Alpha Centauri we would have to wait four years for the light showing our departure to catch us up. We would arrive before we had seemed to set off, just as in Sagan's *Contact*.

Thorne first proposed the use of wormholes to build a craft in a sober paper for *Physical Review Letters* in 1988. He happened to mention that some advanced alien civilization capable of manufacturing such a device to journey between the stars would, possibly as an unintentional by-product, have also built a time machine that would enable them to visit the past.

The popular media immediately saw this as a scientist planning to build a time machine and ran with the story. Thorne never regarded this as the main purpose of his theory, however. Afterwards he was much more temperate in his use of language when writing on the subject in order to avoid being labeled a time machine manufacturer!

What Thorne really discovered was that time machines were of great interest to the media and the public. He would not to be the last scientist to enter the race to break the time barrier, only to face the frenzied attention that came with it.[54]

1992
CHRONOLOGY
PROTECTION

None of these developments did much to help Stephen Hawking square the growing number of potential time machines with his own thinking. The closer we seemed to be coming to breaking the time barrier, the deeper his frustration grew. Though some once skeptical scientists were beginning to accept the inevitability of time travel, Hawking was not so easily persuaded and led the rearguard defense of the principle that cause always had to precede effect. The question was—how could Hawking convince his colleagues otherwise?

Hawking proposed a way that time travel might be blocked from reaching practical fulfillment thanks to what was then an unknown rule of the universe. Once it was found, then it would show why time travel could never happen, even though all indications seemed to suggest otherwise. Hawking called this mysterious property of nature "chronology protection."

This idea was just conjecture, he admitted, not proven fact, but must exist in the scheme of things if we are to stop wormholes from causing mayhem all over the universe.[55] If the chronology protection rule doesn't exist, then countless advanced civilizations ought to be roaming the universe and causing temporal interference to suit their needs. Indeed, a sufficiently ruthless species could simply have used these time tunnels to eliminate all opposition to their domination and be in total command of the cosmos.

Clearly there was no sign of any such time travelers, which further demonstrated to Hawking that some kind of blocking strategy was probably at work in nature to effectively bar time travel from happening. And many scientists agreed. Even in the early 1990s the prospect of someone building a time machine and upsetting centuries of thinking about the laws of nature was still anathema. Heartened by this new idea, a number of researchers actively set out to prove that Hawking's negative approach was correct.

This introduced a growing battle between those who were eager to make time travel work and scientists sure that Hawking's chronology protection law must be out there waiting to be discovered. Jerome Gauntlett of Queen Mary's College in London quickly challenged Hawking's speculation, when he calculated what the universe would be like if there were five special dimensions instead of the four that Klein had revealed in 1926. This additional dimension provided useful resolutions to complex mathematical problems, but as a side effect quadrupled the versions of the universe in which time travel would be a common occurrence. If Gauntlett was right, then time travel was the rule, not the exception.

Unwilling to accept this result, Peter Horava at Berkeley in California probed deeper and suggested a way to overcome it. His argument basically involves seeing reality as a holographic projection—with our limited dimensional senses just observing a reflection of this higher fifth-dimensional reality. This inherent limitation in our perception of reality creates a number of illusions, which are nothing more than reflections of the full picture rather than the full picture itself.[56]

It's possible to visualize this by recalling that a shadow on a wall is a two-dimensional projection onto a flat surface of what is in truth a three-dimensional object. Using this knowledge, it's easy to create a very credible illusion, for instance using your fingers to form a shadow of a rabbit on the wall. Although we see the rabbit in two dimensions, we can easily understand how our senses were fooled when we see the fingers in three dimensions. Without knowledge of the third spatial dimension, we see only a part of reality—not all of it—when we look at the shadow.

Similarly, Horava argues, we see time travel occurring in our three-dimensional perspective of the universe because what we witness is just a shadow on the wall of the extra-dimensional cosmos. If we could see the entire five-dimensional picture, then we would know that time travel is an illusion and does not really happen.

It was a clever way out, but nobody could prove it.

Lisa Dyson at MIT in Cambridge, Massachusetts, came closer still to being both hero and villain when she worked on the problem, designed a time machine, and then found that it seemed to have chronology protection built in. Her concept involved the fifth-dimensional space-time that Gauntlett argued as the best solution for the evidence available. Dyson then calculated how spinning black holes, known already to be a plausible way to create time travel, would behave in this fifth-dimensional reality. What she discovered was that they turned into excellent time machines by creating gravity waves as a sort of fuel to drive the black hole into an ever-increasing spinning motion. This fueling process meant that we didn't need to power the rotation ourselves with unfeasible levels of energy—an apparently damning reason why such a device might never be built. Her version was a self-propelled time machine. Dyson's work seemed to give further credence to the notion that time machines were relatively simple to construct.[57]

But a close look at the mathematics of the situation re-vealed otherwise. Once the rate of spin of this black hole time machine reaches the level at which the jump-through time should occur, something stops it from going any faster and time travel simply never happens. That something is a complicated interaction in terms of subatomic particles, which seemed impossible to overcome.

Was this proof of chronology protection at work? Not quite. Dyson's discovery was specific to a fifth-dimensional spinning black hole. It could not be applied generally. Be-sides, the chronology protection might only be a problem for this one device. There could be other types of time ma-chines that would work even in five dimensions, but with-out exploring them, who could say?

Perhaps Deborah Konkowski. She seemed to hold the answer to Hawking's prayers. Indeed, her early research had probably set Hawking to thinking about chronology pro-tection in the first place. Working at the United States Naval Academy, Konkowski calculated what would happen with energy fields acting in a vacuum in connection with pho-tons of light. If a time machine were switched on near these energy fields, she discovered, then the photons within the machine would begin to increase in size, apparently with-out upper limit, creating an overload effect. This would stuff excess energy back into the balance of quantum reac-tions taking place as the device began to operate, prevent-ing the time machine from working properly, if at all.

Hawking was delighted to hear this, but found a fatal flaw. To prevent the time machine from working completely, he realized that the nature of the space-time surrounding the time traveler would need to oscillate—switching from one dimension to another very rapidly indeed. Unfortunately, the results of his mathematical study of this possibility showed that this was not necessarily always true. The time machine might still just about succeed in overriding the energy overload and get under way before the critical limit was reached. At least under these circumstances, the chronology protection rule did not hold.[58]

Nobody has yet demonstrated a universally applicable chronology protection rule, but it's not for lack of trying. While Hawking wonders if it will ever be found, he is rather more philosophical about the possibility of a time machine these days.

Prospective time travelers have worked to find a way around this issue. Russian scientist Igor Novikov, with enthusiastic support from wormhole time machine designer Kip Thorne, has attempted to do just that. It began with a paper that Novikov prepared for the theoretical physics department at the Nordic Institute in Copenhagen. He based his work on the principle of minimal action, which has been known about since the time of Newton. The principle demonstrates that the course that follows from an event tends to be the one that involves minimal action, the least energy expenditure, or the least time. It can be summed up as "nature is lazy and takes the easy way out."

A ray of light could go from A to B via numerous routes but is most likely to go in straight lines because this is the path of "least time." But when passing through some other medium, such as glass, light curves because the path of least time now requires it to move along a different track. Novikov was able to show that the same rule applies to photons of light in time travel situations.[59]

Thorne then applied the law of self-consistency to time machines following Novikov's path of minimal action, further restricting what light could and could not do. The law states that any change we make must always be defined by the laws of nature, our own physical limitations, and the restrictions imposed by the structure of reality. A perpetual motion machine, for instance, could never be built, however clever we become, because it requires that energy be created out of nowhere, whereas energy, in all of its forms—including that bound up inside matter—has to remain constant in any system. So if we are barred from doing some things no matter how hard we try, then even if we can time travel into the past we cannot force any chaotic changes when we arrive. This eliminates the headache over temporal paradoxes that have so obsessed many physicists.

The law of self-consistency will allow time travel by way of the rules of physics, but it limits what you can do with this ability. You can go into the past in a time machine, provided the correct historical flow of major events is maintained. Minor details might be altered but not anything that has a lasting impact on the future state of the universe. It is as if there is a sort of improbability threshold that, once passed, involves the expenditure of too much energy to allow very unlikely changes. You could not, for instance, go back and prevent your own birth. This event is far too improbable and so gets protected by the laws of physics. But you might be able to alter the precise date of your birth.[60]

Science fiction, of course, had already paved the way for such considerations. In the award-winning episode of the TV series *Star Trek* called "City on the Edge of Forever," Captain James T. Kirk visits twentieth-century America and falls in love with a young woman. When they discover that she will die in a car accident soon after their arrival during the era of gangsters and Prohibition, the crew gallantly considers rescuing her from this historical fate.

But with the cold logic of temporal reason, they discover that if she is rescued, many terrible things would happen that never did occur because she died. Change one event and the future becomes a pile of dominoes toppling over. Kirk, faced with this terrible dilemma, reluctantly chooses to allow history to unfold as it should, but the story implies that time may be sufficiently flexible to allow free choice as to whether to change the past or not.

Other science fiction stories have suggested the opposite

conclusion—that time is like an elastic band and will always snap back to maintain its correct shape no matter how hard you try to change things.

We may discover which conclusion is correct only when we make that first journey in a time machine.

1994
STRINGS ATTACHED

The trouble with all of the time machine concepts on the drawing boards in the early 1990s was that nobody had actually built any of them. Most were considered feasible by a growing number of physicists but required some technology or another that was beyond our capacity to achieve at that time.

Scientists are a patient bunch. They were happy to wait until we had the ability to test chronology protection (which would deny the very act of time traveling) or the principle of self-consistency (which would allow travel into the past but restrict any changes that we could make once we got there). Either or neither might be vindicated in the face of a real attempt to break the time barrier. But humanity has always had its fair share of gung-ho adventurers who won't wait if something seems even remotely possible. By the 1990s these entrepreneurs were selling their designs for homemade time machines on the Web, urging an army of would-be chrononauts to report on the results of their time travel experiments.

Unfortunately, do-it-yourself time machines often have strings attached. Time traveler Steven Gibbs provided one of the more bizarre stories, as he seems to have built one of these devices after using it to go back in time and make contact with himself!

Gibbs moved to Norfolk, Virginia, in 1981 and within hours of arrival received a letter dated in the future he had allegedly written himself. It described the work that he would be doing during a long battle to achieve time travel. Naturally enough, Gibbs thought the letter was a practical joke, but it still piqued his interest—especially when things started to work out just as the letter said they would.

In 1994 Gibbs was taking this daft tale in stride. When the date arrived, the moment when he ought to have sent the letter back through time, he made sure that he did not do so. But by then he had developed a time machine, just as the letter predicted, so he could never be sure if the initial letter was simply a friend's idea of a joke that inspired this achievement or if an alternative version of himself from some parallel universe had actually sent it.[61]

Gibbs calls his device a "hyper-dimensional resonator." His design sketches show a large electromagnet, a coil rather like the one designed by Tesla, and a quartz crystal vibrating at 8.7 MHz to create a resonating energy field. Though the machine was to many observers more likely mumbo jumbo than good science, Gibbs claimed that it worked for him and for others also.

Gibbs sold his machines on the Internet for $360 and hundreds of keen chrononauts signed up, despite Gibbs's warning that he could not guarantee that the device would work for them. He even warned that they might get trapped in the past![62]

I once came close to stepping into such a time machine. It was a machine built in a London garage by chrononaut Tony Bassett. This device, he claimed, induces an altered state of consciousness that allows the user's brain to be projected into another time. A fascinating man and certainly dedicated, Bassett sells his machine, which he calls a "bio-energizer," for less than $500 and offers a money-back guarantee because people respond to his time machine in many different ways. If it is not right for you, he explains, then this is neither your fault, nor the fault of the device, and he would rather sell these things to people who can really make use of them.[63]

So what is a bio-energizer? It seems to have a lot to do with personal sensitivity to EM radiation. A clue as to whether you are susceptible and might find Bassett's machine successful is to gauge whether you react to the presence of an imminent thunderstorm. If you do, feeling a tingling sensation in your head, a sense of oppression from above, a migraine-like headache, or even light-headedness, then you might qualify as a Bassett chrononaut. About one in 10 people show such symptoms during the onset of a storm; they are hypersensitive to the intense electrostatic charge and ionization in the atmosphere that thunderstorms bring. Since his machine duplicates this situation, individuals who experience such things are most likely to respond to his device.

Bassett built the first device in an archway under a railroad bridge in Chalk Farm, London. He originally intended to use his "thunderstorm in a box" to boost the immune systems of cancer patients. Tests have revealed that Bassett's machine, a box approximately a foot square, uses a powerful magnet and electrical field to generate a broad range of high-frequency energy, including a strong magnetic field not unlike that generated by CAT scanners developed for medical diagnosis. The air ionizes within a several foot radius of the device, but the effects are strongest the closer you are to the box. Those using it for experiments tend to lie down with the box touching their heads. Another person, acting as a guide, can watch for signs of any ill effects and keep the test subject's mind anchored, as there is a tendency to drift away into such a dreamlike state that the participant might start to wander incoherently.

Most participants feel only a mild tingling sensation. But successful "travelers" feel disembodied, as if they are drifting upwards towards the ceiling and once there they can direct themselves to any place or time. It's arguable whether these sensations and subsequent experiences represent actual voyages since most reports can never be checked for accuracy. And Bassett agrees that some participants feel like they had been dreaming and accepts that this may be all that there is to it. However, he says that it has sometimes been possible to look into the events reported during experiments and those cases prove that consciousness truly does locate itself in another time and place.

The machine's time travel potential was revealed in the early 1990s, when some participants described finding themselves in the past, witnessing, for instance, a scene from the French Revolution. Others noted that when they "projected" they became aware that there were many other dimensions. When they focused on the structure of subatomic matter, they could see a network grid of interlocking timelines.

While viewing other dimensions, several witnesses began reporting that they could see a thin line like a string stretching towards infinity at either end. Move along the string one way and you enter the past; move the other way and you visit the future. Only a small percentage of users claimed to be able to identify this "time dimension" and travel through it, however.

Bassett claims that some scientists have bought versions of his machine. One machine, he says, sits in a London hospital but he has been asked not to identify the location because those involved feel that its presence would not aid their reputations as responsible researchers. Obviously Bassett's work needs proper scientific evaluation if it is to be taken seriously as evidence for time travel.

Experiments to verify the time travel potential of this machine have been limited. Relocation tests include a woman projected into a nearby street to note the details of cars parked there. An experimenter went to that street and confirmed that these vehicles were indeed present and that they matched in the details, such as color and license plate. But in other instances the cars seen were not present at the time of the experiment and yet were present when the area was checked again later that day. It was as if the machine had projected the subject into the street some hours into the future. On the other hand, the subjects may have seen these cars in the street before and stored data about them subconsciously.

I sat in the presence of the bio-energizer for two hours during one such experiment. The machine generates a strong static field. At close proximity there is a tingling sensation and noticeable sound, a cross between a buzzing and a humming. I could see how it might induce altered states, given its soporific effect, but it only gave me a mild headache. I did not travel through time, but Bassett seemed sincere.

Bassett's experiments match some fascinating research conducted by highly regarded neuroscientist Michael Persinger at Laurentian University in Sudbury, Ontario. He has been conducting tests on subjects for 20 years and has documented his findings extensively in the medical and scientific literature. Persinger designed a machine that generates a strong EM field to stimulate the brain. It triggers an altered state of consciousness and precipitates visions in many of his test subjects. These often involve feeling a sense of timelessness, disembodiment, and many of the features recorded by Bassett's own subjects. Persinger is researching what he believes to be natural pockets of free-floating energy generated by the Earth's geology and calls them "transients." He has designed his machine to simulate this energy to test his belief that these naturally occurring EM fields affect suitably sensitive people who happen to be in the wrong place at the wrong time. The result is a set of physical sensations or visions of weird things that they interpret either as time slips or alien contacts.[64]

Apart from such research from Gibbs, Bassett, and Persinger, who all come from very different perspectives and a wide range of scientific backgrounds, it looked by 1994 as if no humans were going to be time traveling backwards through time in a more demonstrable fashion. And then Richard Gott of Princeton, who had already come up with the concept of the time mirror, stepped up to the plate again. He had not given up on finding a practical means of time travel.

Gott believed that the answer lay in the rapidly escalating search for the TOE (Theory of Everything). String theory offered the best hope for such a theory. Strings are thought to be the most basic element of the universe at a fantastically minute scale, far smaller than even the tiniest subatomic particle. They are the building blocks out of which matter and energy is created, and these strings are what make up subatomic particles. Being so very small, subatomic particles are shielded forever from direct human observation, according to Heisenberg's uncertainty principle.

In order for strings to exist, multiple dimensions (up to 10 of them) are necessary, mostly new dimensions of space but with one extra dimension in which time would act quite differently. These new dimensions would have to be curled up so small that they hide inside the tiniest bits of matter that we can observe.[65]

String theory reveals a universe where time travel may be commonplace in order to produce the apparent chaos that we experience within subatomic space. Gott, therefore, began to calculate how we might build a machine to help us travel into the future or the past. It would do so by manipulating naturally occurring events taking place all around us but out of sight.

Gott soon found that when two cosmic strings interact they effectively become natural time machines. They distort the flow of time so much in their immediate vicinity that they create eddy currents. Most scientists accept that there will be two types of string at the heart of all things—super-strings, which have zero width and are bound into incredibly small loops, and cosmic strings, which do have width, although this is almost imperceptibly narrow. Amazingly, cosmic strings can be thousands of light years in length. In some theories, strings may even have infinite length, although usually they are imagined as looping back and forming rings with enormous diameters.

Even with hardly any width, strings would have enormous mass because of their vast length. Huge mass creates a big gravity well that bends space and time in a manner that can be precisely calculated. Consequently, a cosmic string positioned in the gap between the Earth and a very bright object far away in space would distort the space around it into what Gott proved would be a conical shape. Any light rays heading towards Earth from that distant bright object would curve around the cone into two paths, one on either side of the warp in space that the string creates.

Gott became very excited when he realized that this discovery provided him with another way to travel into the past. And it just might be the simplest way. One of these light rays must follow a path considerably shorter than the other because it is "bent" less on its voyage towards Earth. Because light travels through the vacuum of space at a constant speed, then these two light rays may set off at the same moment but will not arrive together. One will travel through less distance thanks to the presence of the cosmic string. However, it has still traveled at the same speed as the other ray of light despite reaching its destination early. Although we cannot view the motion of this light ray through those dimensions hidden from us, it seems to our senses that one of these rays has traveled faster than light. That speed is an illusion caused by the shortcut, but explains how it can arrive too early. It will travel so fast, it will experience time travel and have actually moved into the past.[66]

Unfortunately, we cannot detect cosmic strings. But we can look for the examples of apparent backwards time travel they may produce. Gott saw a good way to test this principle in action thanks to quasars. These quasi-stellar objects emit so much energy that they resemble stars when seen from Earth but are actually masses of gas being sucked inexorably towards the pull of a black hole. They are found deep in outer space but are so bright that their rays of light easily reach us. The path of any light from a quasar towards the Earth would be split should a cosmic string be located in between. Even if we cannot detect this string we could look for this behavior of the light if it is divided into two separate paths on route to us. One of these routes will be through hidden dimensions and so result in a shortcut. To our eyes it will appear to travel faster than light and move backwards through time.

Gott tried to show that it was theoretically possible to construct time machines using cosmic strings by investigating quasar 0957, which lies almost 9 billion light years from Earth and beyond a galaxy that stands in between it and the Earth. This galaxy has billions of stars and enormous mass, meaning that it distorts space and bends the light passing through it from the quasar into a cone—just as an adjacent cosmic string would do. From Earth we should therefore see twin images of the quasar, one on each side of the galaxy, formed from the two paths that the light rays must take as they head for Earth. This is what Gott's team at Princeton searched for.

This effect they were looking for is quite real and is known as gravitational lensing because the gravity well of such a massive object acts like a lens to focus light rays down split paths converging on the Earth. Since this particular quasar has a varying pulse of light, it gave Gott's team the opportunity to check whether the twin images arrive at our planet at different times. Gott's team spotted the first characteristic change in emission from one of the twin images, then just over a year later spotted the second pulse. This experiment proved that the first light beam from the quasar reached the Earth faster than the second after taking a shorter path through space-time to get here. That first light ray traveled backwards in time by approximately a year. However, the single year saved on this trip was out of a total journey that had lasted nine billion years.

None of this proves the existence of cosmic strings, but it does show that the principle is correct and illustrates that if they exist, they would distort space and time, as Gott predicts. But if we have to rely on quasars to provide us with a temporal shortcut, the huge distances involved would make strings of little use for our purposes. Nobody will want to time travel back in time by a year if they have to fly for nine billion light years to achieve it. But cosmic strings, if they exist, must permeate the universe. If so, they could be found in our immediate vicinity, not just far out in space, making them far more valuable than black holes or wormholes as the basis for time travel.

If cosmic strings do exist across the cosmos, then there should be an interlocking system of time travel routes across the universe. Go one way and you will travel back a year, go another and you might reach back centuries.

Gott calculated that the best mechanism for time travel would occur when two cosmic strings chance to pass each other as they "wiggle" through dimensional space—something they may well do at very great speed. He found that this event could happen commonly if cosmic strings abound, as theory predicts, and when it did so would produce extraordinary consequences. If the passing strings were located between two planets, it would be possible to use their side effects as a time machine. We could utilize the massive temporal distortions they would introduce to leave one planet, fly to the other, return back to the first one, and greet yourself before your departure. Cause and effect would then be right back in the melting pot. You should be able to physically prevent yourself from setting off on the mission when you intercept yourself after coming back. But if you do meet yourself and stop the journey, then you must have made the journey in order to be there to stop yourself making it. You see the problem!

Gott's work created a ripple of excitement within physics, because it offered what was then the most practical method of travel into the past. No fantastic amount of energy was required and the speeds involved are not faster than light speed or necessarily very close to it. So this kind of time travel is perfectly feasible. If and when cosmic strings are identified, Gott's time machine should work. However, a familiar problem has reared its head: Even cosmic strings turn out to have limits.

The laws of physics dictate that it will be possible to visit some events in the past, but not others. Caltech physicist Curt Cutler calculated the extent of these restrictions and defined the "Cauchy Horizon," named after a French math whiz whose nineteenth-century equations were used in the research. The Cauchy Horizon dictates that there must be a zone of useful action around every time machine built using cosmic strings.[67]

Enter the Cauchy Horizon, and time travel into the past becomes possible. Exit that zone and the time machine stops working. A device based on passing cosmic strings will form a buffer zone around these strings dictated by how close they are situated together. Only at their closest proximity will the time machine work. This is probably why we are not surrounded by time machines using cosmic strings flitting around the universe from some point in the future. People in our future may well be able to travel backwards in time this way, but only within the limits dictated by the Cauchy Horizon of their time machine. And that might not include the present day.

Even so, the possibility exists that we can enter a time machine's buffer zone at any moment. When this happens, time travelers could suddenly appear in our midst, even years before they will build and switch on their time machine. Indeed, the first clue that Gott's device will work could be a future, elderly Gott arriving to tell us that he is from 2025 and has traversed the Cauchy Horizon using a device that he has yet to perfect! Steven Gibbs's claims might not seem quite so fanciful in light of such revelations.

Currently we cannot transport a human being at sufficiently high speed for such a time machine to work. But there is another way in which Gott's research might be more readily put into practice. Accelerating large things like human beings requires great energy, but it is much easier to accelerate subatomic particles as these are very light, and indeed we have already done this in particle accelerators to the degree necessary to utilize cosmic strings for time travel.

It seems likely that the first practical application of this type of time machine would involve sending some kind of signal back through time in the form of subatomic particles. The first news of successful time travel may well arrive in code from one scientist to another informing them how to perfect their research—a high-tech version of Gibbs's letter. But a real memo from the future will not come through the United States mail. It is more likely that our computer screens will bring the news of the first successful time travel experiment. And we may hear the results before we even switch it on!

1996
CHEATING GRAVITY

One of the strangest days in my life occurred when I sat in the office of Professor Eric Laithwaite, a renowned scientist at Imperial College in London, with a man who professed that he had a message from an alien. The man, who was a Jell-O manufacturer, explained various supposed details about how spaceships flew using antigravity, something he had learned from a beam of light he experienced in his garden as a young child. Laithwaite listened politely, raised eyebrows a few times, made copious notes of what was said, but gave little indication as to whether he believed a word of this or not. Then he asked if we wanted to see a demonstration of antigravity.

Laithwaite was an engineer and had designed a linear induction motor that allowed trains to ride on a cushion of magnetic air. He floated spinning gyroscopes for us in this way, producing what appeared to be an antigravity force. These would later become the basis for some amusing children's toys that cause small objects, such as a coin, to float in space. I was suitably impressed with the demonstration, especially as the professor had grand aims. He wanted to take his experiments into orbit and test a gyroscopic antigravity motor free of the problems induced by the Earth's own gravity well. Sadly, Laithwaite never saw his dream fulfilled, though he tried.

While reeling off various ideas about how to create antigravity, the Jell-O maker told Laithwaite that he knew the professor was stuck on a problem but that he could provide him with a solution. The scientist looked briefly startled.

Laithwaite later admitted to me that this was not the first time he had been approached with information from such a weird source. In fact, he said, he had once been approached on the street by a person who gave him clues that proved vital to his gyroscope experiments. That stranger did so after claiming to have crossed the barrier of space-time in order to visit the professor. Laithwaite had no idea if this claim of time travel was true, but this unknown man's words, whatever their origin, proved influential. This explains Laithwaite's willingness to meet the

Jell-O manufacturer with the antigravity ideas, though, as I promised Laithwaite, I would not recount this episode until after his death. I will never know whether this celebrated electrical engineer meant what he said that day or was testing the limits of my gullibility.[68]

For many years I pondered the meaning of that surreal day in a scientist's office before it took on a whole new meaning when another scientist claimed to have discovered how to conduct antigravity, which may be the missing link in the race to build a time machine.

You might be wondering why antigravity is necessary for some types of time machine. In order to travel back in time, you need something to prop open the wormhole mouth— like a stick holding open the jaws of an alligator—while you dare to peek inside.

Exotic matter is required because this can create the localized antigravity needed to hold open the wormhole. Otherwise, it inevitably collapses under the huge gravitational forces surrounding its mouth. If something could counteract those forces—creating a patch of antigravity— then it might be possible to keep that wormhole afloat long enough to time travel.

By 1996 no suitable form of exotic matter was on the horizon. Even when condensates were produced that could slow the flow of light, they did not generate any useful antigravity. If Laithwaite had succeeded in making real antigravity forces that could be controlled, then any time machine using them to navigate a wormhole might be able to clear a path ahead of itself.

While antigravity seems like science fiction and is often used in that context, it has a sound scientific footing. As long ago as 1948, the Dutch physicist Hendrik Casimir demonstrated a form of antigravity. What came to be known as the Casimir Effect set up the idea of using anti-gravity to prop open wormholes.

The Casimir Effect depends upon strange things that take place in the gap between two electrically conducting metal plates brought close together in a vacuum. The vac-uum, far from being empty space, is in fact seething with energy waves that create and annihilate countless particles one after the other. Because both normal particles (adding positive numbers to the energy balance) and antiparticles (adding negative numbers) are formed inside the space be-tween the plates, the region is filled with energy with an overall negative balance. The space outside the plates lacks this high level of antiparticles and as a result does not have this negative balance. This means that there is a net positive pressure coming from outside the plates and it squeezes them together to create what is, in essence, a region of neg-ative gravity when the device is switched on, and electrical conductance occurs.

Granted, having antigravity between these plates has no practical value. And not until 1997 could the experiment be done with any precision. But it did open up the possibility of using such an effect to control wormholes. Or, better still, to find an easier way to produce antigravity effects without the need of a quantum vacuum.

Laithwaite had been trying to do just that, after finding out that you could use superconductors instead of ordinary conducting metal plates as in the Casimir Effect. In his experiments, he placed his spinning gyroscopes above a superconductor and they floated merrily in thin air, seeming to defy gravity in the process. But what is so magical about a superconductor?

All things conduct electricity to some degree, although their resistance to its flow may vary. Water, for instance, conducts very well, whereas many plastics do not. The existence of superconductors was found as far back as 1911— and, as the name suggests, these are basically any substance that allows electrical conductance to occur virtually unimpeded. A number of metals become superconductors when cooled down close to absolute zero, or minus 273 degrees centigrade.

For many years this was a phenomenon without much application, because cooling something down to such temperatures is extremely hard to do—the reason why it took so long to create a Bose-Einstein condensate and "freeze" light in its tracks. There were many attempts to find substances that exhibited superconductivity but at more manageable higher temperatures. But by 1996 Japanese scientists had found a compound that exhibited superconductivity as "high" as minus 100 degrees, and a Los Alamos laboratory had perfected a film that needed to be at minus 196 degrees but could just about be used in certain applications.

If practical high temperature superconductors can ever be created, they would allow a virtual free flow of current without the loss of energy. That's why superconductivity research is such a big deal. Tampere University in Finland was one place doing such work when physicist Yevgeny (Eugene) Podkletnov made a chance discovery that provoked an uproar in scientific circles.

His experiment used a ring of ceramic material cooled to very low temperatures by a bath of liquid nitrogen in an attempt to induce its superconductive properties. The ring floated on a magnetic field created by linear motors similar to those that Laithwaite had developed. On this cushion of air, the ring was able to spin rapidly and freely, at 5,000 revolutions per minute.

According to Podkletnov, what happened next was a fluke. One of the team was a pipe smoker and as he leaned over the device his pipe smoke drifted towards the ceiling, despite the cold air that hung above the bath of liquid nitrogen. The smoke literally defied gravity.[69]

Tests followed and many further experiments revealed the remarkable truth. Gravity was being reduced in the area above the floating superconductor. It was not antigravity as such, and the physicist never suggested that it was, but this is the story that took root. Such an allegation created rip-

ples of discontent at the university, especially as Podkletnov's research was not a university project; he was simply
using the university laboratory.

Grant funding agencies were soon expressing disquiet at
the wave of publicity about Podkletnov and his antigravity
machine. Although the results were accepted for publication
in a leading physics journal in the United Kingdom, this decision was rapidly overturned when it became clear that
sources at the university were unhappy with the data. Indeed,
some of those involved declined to support Podkletnov's
paper, and others—including the physicist himself—apparently left their posts following the media quagmire.

Not until March 2000, when Podkletnov attended a conference in Sheffield, England, did the real story emerge. Results showed that there was indeed a slight fall in the gravity
well created by the hovering superconductor. This was the
equivalent of a reduction in the mass of about two percent.
It was minimal, and the outcome was not antigravity—
just reduced gravity. Nevertheless the effect was real and was
recorded in many different ways, for instance, by showing a
lowered atmospheric pressure in the region directly above
the superconducting ceramic. In fact, effects were recorded
on every floor of the university building immediately above
the site of the experiment, suggesting that this reduced gravity effect extended upwards vertically for some distance.

The reduction in mass was measured by suspending objects on a balance within the zone where these reduced

gravity effects were recorded. However, a team at the University of London criticized the method, alleging that when they tried to reproduce the experiment they found flaws in the way the measurements were taken.[70]

No longer welcome at Tampere University, Podkletnov was forced to return to his native Russia to continue his experiments. He remained adamant that the effect was real and reproducible. In fact, between 1996 and 2000, he reduced the mass up to five percent. These results were in keeping with relativity physics. Einstein had predicted that a rotating object should reduce gravity in a local sense by creating a similar reduction in mass. However, at modest rotation speeds, such as 5,000 revs, the effect should normally be so tiny as to be impossible to measure in the way Podkletnov had done it. It should certainly not be of the order of two percent, let alone five percent.

One of the teams that tried to replicate the Finnish experiments was based at the University of Alabama. Dr. Ming Li argued that superconductors might magnify the mass reduction effect. But Podkletnov's research proved disappointingly difficult for others to verify.

At present, antigravity research of this kind is even more contentious than time travel itself. Some think that Podkletnov is just plain wrong, while others hope he may have hit upon a way to reduce the mass and gravity of any object in motion. This would eventually allow us to minimize the gravity effects that collapse a wormhole and to power a time machine by prying open the wormhole's mouth long enough for time travel to be possible.

If we succeed in doing this, then travel into the past will force us to confront the nightmare scenario that physics has long dreaded. If we can visit the past, can we alter the past? And if we alter the past, what happens to the world that we have left? Will we return to the present day to find that things are just the same as when we left them, or will we find that we have altered the universe of today in a myriad of different ways just by swatting a fly on a journey into the past?

1997
MANY WORLDS, MANY TIMELINES

Science fiction would be much poorer if not for the concept of time travel. On several occasions great writers have used novels to suggest how we might advance actual research in this field. Indeed, many fine science fiction writers (such as Benford and Hoyle) are scientists whose real life work embraces their fiction. One key success for science fiction came by ending the nightmare posed by the temporal paradox. Could science fiction's clever suggestion for an escape clause work in genuine time travel?

The way out of the dilemma that writers of time travel stories have adopted is the notion of "alternative universes." On this basis trillions of versions of our world might exist.

Go back in time and kill your parents and there is no paradox to consider. Not if we argue that you commit this act in one of the countless similar, but not identical, universes rather than the one where you are born and travel back into its alternative past and do the dirty deed.

If this switching of universes accompanies any time trip, then the "you" that is never born is not the same "you" who prevents this from happening, merely a "twin you" who was never going to be born in the first place. You may even have fulfilled time's destiny by visiting a parallel version of your universe and bringing about a series of events that were bound to happen there anyway.

Fred Hoyle's son, Trevor, followed his father's footsteps into science fiction and adopted some of his ideas to create a series of adventure tales beginning with *Seeking the Mythical Future*. In this saga Trevor Hoyle imagines a series of alternative timelines that coexist into perpetuity side by side. The worlds in this novel are the consequence of persistent interference over the eons by travelers reaching the past from the future. Repeatedly going back in time and altering things has triggered a cosmos that is littered with alternative versions of the original timeline. Whenever someone changes the past, it is not their past that they revise but the past of another universe. In this way time traveling has rapidly filled up the cosmos with limitless variations on a theme.

In one universe, the Nazis develop the atom bomb and force the surrender of the Allies, which leads to a very different version of the modern world. It takes only one slight alteration in the past chain of events to create this radical alternative version, existing alongside our own past that still survives in its own parallel universe. Hoyle suggests that a trillion different versions of the past might exist "out there" and when we eventually are able to time travel we might visit any one of them instead of the familiar version that we are expecting to reach.[71]

The idea for these parallel versions of space and time did not spring unprovoked from the minds of these story-tellers. The concept was developed from actual experiments in subatomic physics. Dr. David Deutsch, at Oxford, is one of those researching modern physics theory to make time travel a reality, and is typical of those who favor this idea of eternally splitting universes. By avoiding Hawking's hypo-thetical notion of chronology protection, alternative universes may be the key to making time travel work.

The concept gained momentum during the 1970s thanks to quantum physicist John Wheeler, who thought alternative universes were the best way to interpret what physicists were seeing in their labs. Experiments showed that whenever multiple possible outcomes can result from subatomic reactions then somehow all of them are real in some sense until something causes just one single reality to crystallize. In light of this finding, the science fiction notion of "many worlds" suddenly took on new meaning.

But the rather mind-boggling concept has some worrying implications. It sees endless alternative versions of this universe, many of which contain alternative versions of both you and other people. This seems rather difficult to believe and only started to be taken seriously by many physicists when fiction revealed how smoothly the concept managed to avoid those horrifying results of building a time machine, visiting the past, and making catastrophic changes to the timeline we know to be correct.

Deutsch points out that every time two subatomic particles collide, they will either bounce off one another, or one of them will fade out of existence and reappear elsewhere in the universe. But we only see the universe where just one possibility took place. Of course, a vast number of these interactions happen all of the time and countless more will follow in an endless succession across the cosmos.[72]

This theory suggests that every possible consequence of every collision splits reality in multiple ways and we just experience one of them. According to Deutsch, we see one of many possible worlds. Although we cannot normally detect these other worlds, they are all out there. Like a railroad track diverging at a junction, we go one way and take that journey, but many other lines branch off and time travelers can go down any of these lines instead.

The problem with this theory—a fatal flaw, as some scientists see it—is that it produces a seemingly ridiculous plethora of slightly different realities. We are not just talking about a few parallel universes, but an unimaginably huge total that is virtually infinite because such perpetual splitting will have gone on since the universe was born.

There will be versions of reality that are so close to our own that we could slip into one of them and never realize that we had moved. But there will be other versions which are extraordinarily different and that split off from the main line so far back into the past that they have become unrecognizable from the one we inhabit.

The implications are mind-boggling. In some universe the conditions may exist where human life spans are prolonged, even to the point of virtual immortality, for example. If true, then versions of ourselves will live forever in these universes—meaning that some aspect of ourselves will never die. The prospect of such a universe is remote but not impossible.

All of this is of more than academic interest. It could have significant implications for time travel. Often our attempts to create a working time machine depend upon solutions to Einstein's equations that are theoretically plausible but do not apply to the universe we inhabit (think of Godel's rotating version of the cosmos where time travel would be commonplace). If the many-worlds theory proves correct, then these universes cannot just be dismissed because they are not the universe we inhabit. In some plausible versions of the universe, time travel will exist and be a regular occurrence. This is vital, because if time travel involves switching universes during any journey into the past, then a sort of temporal osmosis will operate to leak this success across the cosmos and time travel will become possible everywhere.

There could indeed be countless versions of reality where time travel is an everyday occurrence thanks to how that particular branch of the universe has evolved. In order to time travel we do not need to find some complicated method that changes the rules in our own universe, or strive to find a way around the difficulties imposed. Instead we simply need to shift into the right parallel world to take advantage of the time travel conditions that might apply there.

In short, traveling backwards through time might be a matter of traveling sideways through dimensional space.

1998
QUANTUM FOAM

In the Quantum Imaging Laboratory at Boston University a strange experiment occurred that opened the door to a microscopic version of a time machine. This facility studies the way the tiniest particles interact. Using what amounts to a high-tech version of a pinhole camera, the scientists discovered that it reveals things that are otherwise invisible.

In this device a tiny sliver of light enters a totally darkened chamber. Once the light has entered, it is trapped without means of escape. The photons will still strike anything inside the chamber but are rendered invisible because no light reflected from the objects can escape to be detected by our eyes. It's like light rays that get snared by the massive gravity well of a black hole. Once inside the darkened interior, no image will ever be seen by human eyes. But as the Boston lab showed, modern science can make even invisible things visible again.[73]

Using modern computers that can make extremely complex calculations at high speed, the Boston researchers analyzed what was going on inside the chamber by working out the mathematics of the interactions taking place when the light struck its invisible target. They then assessed the time that it takes for a photon to strike every tiny part of this object. The information that is mathematically encoded represents a series of clues to the shape, size, and position of the object that is otherwise blocked from view. When the computer decodes this data, an image of that thing is created.

This process updates a technique used to make a hologram, a three-dimensional image of any object, such as a vase. Although it is just an image and your hand would pass right through it if were placed in front of you, the vase seems to exist in that space in three dimensions. Unlike a photograph, which is flat and two-dimensional, you can walk around the hologram of a vase and see every angle exactly as if the vase were physically present.

Holograms use two light beams, one fired at the object and the other reflected indirectly via a mirror. The time it takes for these light rays to follow their paths provides all the data needed to locate every feature of the object. The idea was developed twenty years before we were able to make a hologram because we lacked one vital ingredient.

That ingredient was the laser, which was invented in the late 1950s. To record the data precisely, completely coherent beams of light must be focused into a single wavelength. Ordinary light is too spread out. The first lasers enabled us to make holographic images just as theory had long predicted. Time travel may follow suit.

What the computers did in the Boston lab involved an updated technique known as quantum holography. It uses two beams of light interacting inside the chamber and measures the times and distances involved. Computers can then re-create the data into a three-dimensional image of the object hiding within the depths of the chamber. From the information analyzed, a very precise simulation of the invisible object as a holographic model is created. Incredibly, it does not matter what the object hidden in the chamber is, or how undetectable it previously was; quantum holography makes the previously invisible appear before our eyes.

The quantum hologram offers a tool that might now let us probe those invisible worlds and enfolded dimensions within the atom. And these are the realms where time travel is known to occur.

As you delve deeper and deeper beyond the scale of atoms and subatomic particles to the far smaller universe where cosmic strings are believed to be vibrating, nothing solid remains. All is in a state of flux. Reality in the raw is highly uneven in nature. In fact, it is like a sea that is constantly frothing on top—something that Wheeler poetically named the Quantum Foam.

Quantum Foam has remarkable properties. For instance, it is riddled with incredibly tiny gaps. These have the same properties as wormholes that might be found far out in space and that we believe connect all parts of the universe together. Indeed, for all intents and purposes, the microscopic wormholes inside the depths of matter are so incredibly small that there is no chance we will ever be able to observe them. Their dimensions are in fact one billion billion times smaller than the size of an electron—which itself is far too tiny for us to view.

Small they may be, but unimportant they are not. Being wormholes, even microscopic ones, these wriggling tubes do not simply act as bridges between different points in space—they cross the time barrier. Theoretically, at least, they could be used to transfer energy at its basic, tiniest level to reach every point in the history of the universe as well as every nook and cranny of the cosmos.

This research has introduced a completely unexpected way to build a time machine—one that does not need a spacecraft capable of traveling the universe looking for a wormhole or that can travel close to light speed to seek out black holes. The universe of inner space is a great place to go hunting for a time machine.

Time travel seems to be a staple property of the quantum foam. But just how do we use this knowledge to break the time barrier? If these quantum foam wormholes are so very, very small, how can we possibly do anything with them?

Anything sufficiently small that could be sent through the quantum foam will go on a journey through space and time and could reach the past. Imagine a quantum physics version of the classic 1966 science fiction movie, *Fantastic Voyage*. In that movie scientists are miniaturized to the size of body cells and injected into the bloodstream of a patient. They must fight off various monsters on the journey—microscopic in size but now gigantic to them—such as human antibodies, in their quest to reach the brain and try to save a dying man. Our new fantastic voyage on a quantum level would involve a much weirder trip through time and space inside a miniaturized time machine.[74]

In fact there are three ways to potentially make use of quantum foam wormholes. We could miniaturize time travelers, *Fantastic Voyage* style, to the point that they could pass through these very tiny gaps. But this seems to be a very unlikely scenario given the scale involved. Alternatively, we might find a way to expand the size of the quantum foam wormholes until they reach manageable proportions where they could provide us with a gateway-sized wormhole in our laboratory, allowing something to pass through. Again it is not obvious how we might achieve this.

The third option would be to send something very small through this microscopic wormhole, an idea that was pioneered by the TV series *Star Trek Voyager*. Stephen Baxter, adapting a basic outline by Arthur C. Clarke, picked up on these ideas shortly afterwards and has produced a superb vision of where such research may eventually lead. Entitled *The Light of Other Days,* it looks forward to how by the mid-twenty-first century we might develop an invention called the "worm cam." In effect this is an extension of quantum holography that opens up those countless tiny wormholes lurking within the quantum foam but without needing to miniaturize humans to act as time travelers.[75]

Worm cams would bridge the gaps in the quantum foam using very small packets of energy with dimensions akin to a cosmic string. In doing so they could reveal the light at the ends of these tunnels, allowing visions of distant parts of the universe at the press of a button. However, as we know, space travel and time travel go hand in hand. Worm cams would also possess the ability to look back through time. Technically it would be little different to receive images via a quantum wormhole beginning one hundred light years away than to receive light from the local vicinity that was emitted one hundred years ago. In other words, these quantum foam viewers could see the light of other days. This breakthrough would open up all of history to direct observation for everything from academic study to the entertainment industry. You could switch on your TV, tune in the worm cam, and watch any event that ever happened or that is happening anywhere in the universe as it unfolds in real time on your screen.

How realistic is this idea? Discoveries made every day suggest that the idea is not outrageous. Indeed, it may come true faster than even Clarke and Baxter imagined.

Worm cams may be tomorrow's must-have personal technology—the equivalent of an mp3 player in the early years of the twenty-first century.

1999
KAKU'S TIME MACHINE

At the new millennium, a physicist in New York who realized that wormholes offered perhaps the best way to time travel, set out to manufacture one from scratch. His name is Michio Kaku, a professor at the City University of New York, one of the most respected researchers in modern physics, and a key player in the development of superstring theory. In many ways he revived flagging belief in the possibility of a Theory of Everything, binding together gravity and relativity with the quantum mechanics of subatomic space.

This complex field of research is sometimes referred to as "M" theory. The derivation of the name is in dispute, with some saying that M means "multiverse" (as in a cosmos filled with multiple coexisting universes) and others suggesting that it more likely stands for "mystery." But these days it is also often used in the context of M for "Mother," as it is hoped that the theory would prove to be the set of rules that gives birth to every known force and physical phenomenon. There is even a variation called "F" theory, in which the F, of course, represents father!

Kaku dates the whole idea back to 1919 when German physicist Theodor Kaluza wrote to Einstein to suggest the first possible way to unify electromagnetic radiation, with its mathematical rules derived by British mathematician James Clerk Maxwell, and Einstein's own newly published relativity theory. At first Einstein was not persuaded by Kaluza's argument because it would necessitate a fifth dimension. But seven years later the Swede Oskar Klein set out in beautiful detail just how it was necessary to have a multiple dimensional universe in order to unify, at least in theory, all the various forces of physics with space-time and relativity. Einstein became a convert and this farsighted concept of multiple dimensions became known as the Kaluza-Klein multiverse theory.

Many scientists toyed with the theory's implications, which eventually led to Einstein-Rosen bridges and wormholes, but nobody found any other evidence for multiple dimensions. Then Kaku came along and breathed new life into the idea with very tiny things called strings threading the universe together out of view within the coexisting realms of Kaluza-Klein's so-called multiverse. By the 1990s Kaku brought together the many variations of string theory into "M Theory." He sees a whole series of dimensions, each creating a mini universe and individually responsible for some key force of nature. One dimension in the multiverse is where gravity exists, another is the home to light, and so on, all of these linked together via a network of invisible (to us) inter-dimensional bridges. Though we can never view these other realities or the strings that underpin them, these forces are like ripples on a pond caused by an invisible wind. We see the ripples (the forces of physics) but not the wind (the strings vibrating to create them in those invisible dimensions).

Kaku feels that it may be a very long time before we can hope to prove M theory directly because the fundamental difficulties of living in just three of these multiple dimensions deny us a direct experience of the rest. But we could find ways to prove M theory indirectly. We know a lot about the nuclear reactions inside the sun without ever being able to send any human or instrument probe into that deadly environment. We learn what we do by investigating how those nuclear reactions affect other parts of our solar system and deducing from these what their cause must be.[76]

In Switzerland a massive new particle accelerator has been designed that will smash particles into one another, and the plan is to use it to look for super particles that are the direct consequence of extra-dimensional strings that vibrate according to M theory. These super particles are the most basic things that we can physically discover inside the atom. If found, they could help us study how the strings that create them produce the forces of physics from out of other dimensions. Sadly it may be 2010 or later before this device is able to look for these particles with realistic chance of success.

But another possible way to prove M theory involves the analysis of what is called dark matter. We know a great deal about the parts of the cosmos that consists of stars and other luminous phenomena, but they form just 10 percent of what is out there. The other 90 percent is the dark matter. The best guess is that dark matter may be formed out of super particles that act as the bridge between our three dimensions of space and the eight within the multiverse. In the deep space beyond our world there may be countless bridges between these dimensions, any one of which could be used as a time machine.

Kaku's M theory views the universe as a three-dimensional soap bubble that floats in a multi-dimensional universe. All around us, unseen, could be trillions of other universes floating in the same way, each one a unique universe with its own flow of events and time. Indeed, the variation—F theory— allows for an extra dimension of time that may flow in the reverse direction.

Making trips through space and time may not necessarily require moving hundreds of light years from Earth. Our steps could just be sideways. Indeed, M theory suggests that other bubble universes could be as close to our own as a single millimeter but invisible and undetectable because they exist beyond the realm we inhabit. Any lab could contain millions of them and a time machine built in that lab could theoretically travel through them without moving very far in a three-dimensional sense.

Superstrings remain just theory, so until we find and harness them this kind of time travel is beyond us. Until then the best way remains to try to build an artificial wormhole. Sadly, the energy needed would still be more than we can produce artificially; although this does not preclude the possibility that subatomic pockets of exotic matter lie undetected all around us—creating countless invisible wormholes that could distort space and time at suitably close distances. The simplest way to look for them would be to detect their side effects—apparent as distortions of time and space that seem to occur suddenly and spontaneously and may be only transient in nature.

From these starting points Kaku set himself the task of trying to build a time machine. His idea involves two chambers, each one containing twin metal plates parallel to one another. The trick comes in generating a sufficiently huge electromagnetic force that would induce massive electric fields between them. The levels involved would be of the order of the super fields generated by Tesla over a century ago when he was trying to design artificial lightning to power homes. Only Kaku's needs would be greater still. But it is certainly interesting to recall that Tesla claimed to experience a form of time travel during his pioneer experiments with such a gigantic EM field. Does this suggest that Kaku is on the right track?

The metal plates in the Kaku time machine must allow as much of the massive energy field to pass through them as it can. The superconductor being developed for various antigravity experiments may become the key to allow a high enough energy field that opens the door to time travel.

If these conditions can be put in place, the machine must distort space-time in the vicinity in such a way as to create a wormhole that links the two adjacent chambers. The result should be a bridge through space-time, which would hopefully be stabilized by exotic matter generated through the Casimir Effect.

The time machine would have to be primed so that there was a time difference between the two chambers. Physicists could do this using the effects of time dilation. This would be difficult, of course, as it would involve physically transporting one chamber at rapid speed—the longer and faster, the greater the time dilation produced. For serious amounts of time travel we would be talking perhaps 95 percent of light speed and journeys across space, but detectable effects could be produced in simpler ways. For example, micro-sizing the chamber and hitching a ride on particles in a linear accelerator, or attaching it to a cyclotron that is sent to circle in Earth orbit. Because time will pass at a different rate for the moving chamber than it will for the chamber left in the lab, when the two chambers are reunited side by side they will have a fixed time difference between them that will allow them to be used for time trips. The distance you can travel back through time using this machine will match the time difference you have primed into it by virtue of its time dilation journey. A substantial time difference would obviously be difficult to program.

Nevertheless, the basic principle seems simple enough, but Kaku admits that it has numerous obstacles to overcome. He knows that at the moment this kind of time machine probably still cannot be constructed, although it is one that we may well be able to build in the next few years given the will and the budget.

Shortly after Kaku came up with his model for a small wormhole time machine, a new development in Europe increased the likelihood for time travel the Kaku way. Sergei Krasnikov of the Stellar Physics Lab in St. Petersburg identified some microscopic wormholes that are larger than those that comprise the quantum foam. Although still incredibly small they possess a property that other wormholes lack, a property that could make all the difference. They are stable.

They seem to have just the right balance of energy requirements to create a wormhole mouth and then produce exotic matter in sufficient quantity to keep it open—perhaps long enough to make the tunnels that bridge space and create a time travel doorway.

Of course, these tiny wormholes are still too small for any practical transmission of data through time. But if we could enlarge them, then their stability might disappear as the delicate balance of forces is disrupted. These new findings are the first significant evidence that larger-scale wormholes might exist in nature. Perhaps they presage the eventual discovery of some that are large enough to allow human beings to pass through. The news has to encourage the hunt for nature's own time machines, which potentially exist in great profusion within the world inside the atom. They could be very close at hand and finding them may turn out to be the quick road to time travel.

But even a stable wormhole will be put under great strain with a time machine rushing through it. The problem is one we know as feedback. If you have a good sound system at home or have attended a rock concert where massive speakers boom the music across a room, then you may be familiar with the problem. Sometimes the sound waves emitted by the speakers will loop back into the system, interfering with the sound coming out to create a cycle of escalating noise known as feedback. This sound energy converts into an increasing and escalating noise that emerges from the speakers as a piercing howl.

The problem with wormholes is that they behave in much the same way, except a whole range of energy is emitted from the mouth—not just sound. The slightest interference from anything that tries to fly through the wormhole will upset the balance and lead to huge feedback. This will rapidly collapse the wormhole mouth and make it impossible for anyone to go far enough to time travel.

However, one way around this problem is negative feedback. As the name suggests, the idea is to reverse the feedback effect that causes the howling. By canceling out the rising energy that emerges from the speakers, you prevent it from cycling into a loop and allow pure sound to come from a speaker.

A very sophisticated computer program can analyze the sound as it comes from the speaker and instantly produce a wave with the opposite properties. When this is fed back into the system, the positive and negative aspects of these waves cancel each other out. You can tailor the effect so that no feedback loop occurs or, even more remarkably, to eliminate the emitted sound altogether. If the two waves exactly match—one positive and one negative—the overlapped result is no sound and no sound means silence.

The technology needed to make the vast number of rapid calculations to overcome the feedback requires extremely fast computer responses and very sophisticated processing power. Once perfected, the same idea could then be applied to wormholes, should we find one in nature that is both close at hand and large enough for a time machine. Any time traveling device fitted with a feedback analyzer and negative feedback emitter ought to be able to counter the disturbances that will seek to destabilize the wormhole. If operated continuously, this technology should prevent the mouth from collapsing.

Boosted by such new ideas and thinking, Kaku's concept for a wormhole time machine was convincing enough to change Hawking's mind. He has been forced to conclude that it is indeed possible to travel through time. But, he added cautiously, whether it would ever be practical, as opposed to possible, is quite another matter.

Hawking's pessimism is justified, of course. Every step taken toward breaking the time barrier seems to encounter a difficulty that stalls the attempt. But all is far from lost. Advances in physics are occurring at astounding speeds. The hunt for the perfect time machine is alive and kicking. Sooner or later one of these ideas will surely pay off.

2000

FASTER THAN LIGHT

On November 2, 2000, a strange series of postings began to appear on a United States website. The person making the postings, John Titor, claimed he was a time traveler. He had arrived from the year 2036 on a mission to study Earth during a traumatic period in our history. He had journeyed back through thirty-six years without using a wormhole or quantum physics but by traveling faster than light.

In 2004, a new communicator, calling himself Professor Opmmur, appeared, saying that he was from 2039 and had met (or would meet) Titor in 2034 but the pioneer time traveler died (or will die) of the flu in 2038. Apart from the fact that such juggling acts with chronology might bring on a headache, no proof has ever emerged to demonstrate the validity of this yarn. It is pointless debating whether these John Titor messages were just a hoax. Common sense dictates that we err on the side of caution and assume this was no more than an amusing leg-pull. But how can we ever be sure?

The problem is that anyone can avoid direct proof of time travel by invoking the law of minimum interference. But we are bound to regret the lack of warning for a tragedy such as September 11th, which might have saved thousands of lives. As it is the Titor story merely serves to illustrate how time travel captures human interest with its numerous weird social implications.

But what of Titor's suggestion that we can time travel just by moving faster than light? That notion became moribund after Einstein's insistence that it was outlawed by relativity theory. Yet in 2000, there were indications that faster-than-light travel might not be so outrageous after all.

Humanity has made extraordinary strides in terms of travel speeds over a relatively short time. For thousands of years we were limited to the few miles per hour that a horse could gallop. When passenger railroads were first built in the 1830s there were even fears that increased power would lead to the maximum tolerable speed being reached—speculation being that human bodies would disintegrate when moving at 100 mph. In March 2004 NASA tested a supersonic jet using rocket motors that flew at seven times the speed of sound—over 4,000 miles per hour. Spacecraft can now go even faster.

But even speeds of 25,000 mph or more reached by a probe in the frictionless environment of space are useless for a journey to another star system. At that speed it takes years to send camera probes on missions to study "local" planets like Jupiter. To get to the closest stars would still take hundreds of thousands of years. Obviously that would be a bit of a drawback for an episode of *Star Trek*, which is why science fiction writers invented a faster-than-light propulsion device almost as soon as Einstein said that it was impossible!

Of course, very fast speeds are possible in space. Once you give a rocket a boost free of Earth's atmosphere, it could go on accelerating for years. By the end of the twenty-first century we may well have designed ships to take probes to other star systems in a fraction of the time now required by using ionizing radiation converted directly from matter that is sucked in like a vacuum cleaner as the spacecraft moves. As this energy is converted to propulsion it generates continuous fuel. But even if these propulsion systems provide us with enormous speeds, as high as a staggering one *million* miles per hour, all journeys between stars will still take longer than a human life span and many would require hundreds or thousands of years to complete. Even accelerating ships to speeds that are a modest percentage of light speed is well beyond our current reach. Faster-than-light travel just seems impossible.

Travel beyond the speed of light makes the theory of relativity blow a fuse, because it involves putting negative figures into the equations. However, in the 1990s we saw increasingly bizarre experiments with exotic matter and antigravity that proved how the idea of negative energy and reverse gravity can be justified from direct observation. Some physicists began to wonder whether there might also be solutions to relativity equations where the seemingly absurd negative figures introduced by faster-than-light travel would presage something that can be re-created.

The other side of the equation (literally) involves exceeding light speed by somehow utilizing negative energy. If such a concept has any existence outside mathematical abstraction it could provide a way to attain such speeds and, of course, time travel. While this may all seem like nonsense, amazingly, in 2000, physicist Ken Olum at Tufts University in Massachusetts opened the door to making this a reality.

In his experiments Olum designed mathematical models that sent EM radiation along thousands of different paths between two locations in space. He then calculated how this radiation would interact with other particles existing in space as it completed the various paths to make such a journey. On each path the radiation interacted with a range of subatomic particles, thereby creating a vast range of collisions, annihilations, and formation of new particles of the kind seen in particle accelerators. He found that the time taken for each journey depended upon the kind of particles the EM radiation intercepted on route. The fastest path occurred when the beams of EM radiation encountered pockets of "negative energy." His calculations revealed that this odd phenomenon was an occasional result of how the particle interactions would balance out along the various possible paths that the radiation could travel. Olum found that there were indeed regions of space where the billions of subatomic particles that might be met by the radiation possessed such a range of individual properties that the net result of summing their masses together might now and then form a negative number. This negative mass was a statistical consequence but it has very real effects. Any journey that involves quantum interactions has to take account of every possible path that might be taken, so even these statistical flukes make a contribution to the actual course of the EM radiation. This course averages out all the options—including those where negative mass can be encountered, making the effects of this phenomenon directly relevant. Negative mass reduces local gravity, which in turn increases the theoretical maximum speed for travel through that region of space. Olum suggests that by controlling such pathways we might be able to boost speed beyond that of light.[77]

This research might suggest a way to design a Faster-Than-Light drive for a spaceship. You would need to have a propulsion system that cleared a path through particles in space that are given negative mass, so that the craft can be sucked through the bow wave that will be cut ahead at ever-increasing speeds. While still largely speculation, Olum's work at least hints at the possibility of success.

Olum's theoretical breakthrough created further interest in seeking to push back the boundaries of the speed of light and led to some startling experiments with stunning results. In the summer of 2000, several major American newspapers carried a story about a team of scientists at the NEC Research Institute in New Jersey who had created a laser beam that beat the speed of light. But, through no fault of the scientists, some overly eager media misinterpreted the meaning of these results.

The physicists—Lijun Wang, Arthur Dogariu, and Alex Kuzmich—did something that just looked like faster-than-light travel. Working for an electronics research company, the physicists were seeking practical applications from altering the wavelength of EM radiation and were successful, making the shorter waves longer and the longer waves shorter as the beam traveled from its starting point towards its destination.

Normally a beam of light diverges, as with a flashlight, and by the time it shines onto a wall is much larger in area. That happens because of the different wavelengths involved in any ray of light. These wavelengths from normal light strike a target at varying intervals, creating the spread and diffusion that we see with any projected beam. Indeed, single wavelength lasers are useful in many practical applications largely because they do not cause this diffusion. They retain all their energy over long distances and can even be sent the quarter of a million miles to the moon without significant divergence, making them extremely valuable as highly accurate tape measures. Ordinary light could not be beamed to the moon, because by the time it had traveled that far it would be so spread out it would have lost all power and focus.

By manipulating the wavelengths of a beam of light in transit, the NEC experiment made ordinary light more coherent and let it keep more of its power for use over longer distances. The development of holographic (three-dimensional) television will probably only be commercially viable when ordinary light (instead of lasers) can be used in coherent form to transmit the data. Telecommunications agencies are particularly interested in this kind of research for the development of tomorrow's technology.[78]

The NEC experiment was less ambitious than a 3-D TV test. It set out to "channel" light much more efficiently through ultra-thin strands of fiber-optic cables that leak minimal energy. This breakthrough in 2000 was the first trial in a non-opaque medium. In the New Jersey experiment some parts of the light beam traveled faster than others and arrived at the destination earlier. This slightly varied the speed of light within pockets of energy inside the beam. But summed across the whole beam, the total still amounted to sub-light speeds.

A subsequent experiment at Middle Tennessee University produced a partial increase in light speed by up to 400 percent. But again this had no faster-than-light consequences for the beam as a whole. Because of that limitation the experiment would have only rather limited value in time travel. The experiments found that the faster certain waves traveled, the more distorted and weaker they became.

It seems that there is a fundamental cutoff point beyond which no useful energy can be sent faster than light. This amount is below the threshold needed to beam any coherent message back through time. So another possible time machine seemed to have hit a dead end.

Nevertheless, these experiments do show how light speed can vary. It has a fixed speed of about 186,000 miles per second when moving in a vacuum (such as through space). In other mediums, the speed of light will differ though it will always be extremely fast. Light always appears to us to travel instantaneously whatever naturally occurring medium it passes through. We only notice the effects of its speed when it travels huge distances.

In a very real sense the relationship between the speed that light takes to travel and how far it has to go creates the experience that we perceive as time. Various time machine designs have sought to alter the one factor in this equation that appears to be controllable—the distance traveled—by envisaging ways to warp space or use wormholes to take a shortcut. That would allow us to cheat time, even if we stay below light speed.

Altering the other factor in this equation, the speed at which light travels, would change our perception of time. Speeding light up beyond 186,000 miles per second would make events arrive at our senses more rapidly, producing a form of travel into the future. But slowing it down significantly would have the opposite effect. Time would pass more slowly from our perspective, causing us to have to wait longer for images from any event to reach us than they normally do. An event that happens today may only seem to happen next week if light slows down as it travels towards us. We would be able to travel (or at least view) the past.

Science fiction author Bob Shaw suggested what life might be like in a world where we learned to slow down the speed of light. Shaw imagined a special kind of glass that could reduce the transmission of light to a velocity far below 186,000 miles per second. At just a few millimeters per year light would take ages to pass through a glass sheet and emerge out the other side, not the millionths of a second it now takes in our everyday experience.

Shaw's idea of "slow glass"—as a pane—could be put in front of the cataracts at Niagara Falls, left there for a year to store the light rays inside, then sold and installed in your home. When the light rays complete their tardy journey through the pane, perhaps months after they first entered, the window would start to reveal the scene at Niagara as if your window were in front of the falls. It will remain in view through your slow glass windowpane until the moment arrives when you see the glass being taken away from Niagara Falls to your home. Then you can replace it with a new pane with a different vista—perhaps one that has been left on the moon by astronauts.

Bob Shaw saw this invention as more than just neat home furnishing. It was the stimulus for a story in which a sheet of slow glass recorded a murder. The authorities could now just wait for the killer to be revealed when the light of this crime finally emerged from its slow trip through glass that chanced to be near the spot where the homicide had been committed. I doubt that the author realized that science would prove his insight to be strangely prescient.

By 2000 experiments were using the smeared gaseous Bose-Einstein condensate mentioned at the beginning of this book to slow time. This substance can reduce light speed to a crawl as it passes through it, much like the "slow glass" Shaw imagined thirty years ago.

Theoretically, something like Shaw's invention is now very much within our grasp. But while this "slow gas" offers a sort of time travel, it will not shock the system in the way that some time machines might do. It produces results not unlike a video camera. The past would be revealed to us by virtue of the light from long-gone events being held in limbo and then released to our view when we choose to see them. But the very low temperatures necessary to make this happen count against present viability of such a time machine.

But this raises an interesting question: Should we be searching for pockets of natural slow gas that might exist, especially in the coldness of deep space? Floating condensates would slow any light rays passing through them and produce time anomalies. If they existed in our atmosphere, they might trigger sightings of what seem like ghosts, where a replay of some past event is witnessed. It would not be a rerun, as such, just the original light emerging after completing its long-delayed journey through this vapor to appear in the form resembling a phantom out of time.

Of course, the slow gas experiments are not deliberately trying to achieve time travel. Their aim in altering the speed of light is to boost the ability of wires and cables to let energy pass through them with minimal data loss. Electrical signals, for instance, are transmitted through wires at only about two-thirds of light speed and so lose much information on route thanks to the effects of impedance. If these effects can be reduced, then greener energy resources will be possible and even greater computing power achieved. Time distortion is merely a side effect.

Needless to say, if light speed can be slowed to zero in certain mediums during condensate experiments, the opposite might also be true. Perhaps one day we will be able to speed up light beyond 186,000 miles per second. If we can do that, then a time machine will be reality.

2001
HANDSHAKES
FROM THE FUTURE

By the beginning of the twenty-first century, time machine ideas were in abundance but time machines that worked still seemed a long way off. Theoretically they were no longer just possibilities. Huge advances had been made and time travel had certainly left the realms of the corny movie and firmly entered mainstream physics. But being possible was not the same thing as being practical, as Hawking had reminded.

For John Friedman, a physicist at the University of Wisconsin in Milwaukee, the idea of building an H.G. Wells style device to carry a human being into the past was a nonstarter. Faster-than-light travel could never be successful enough to allow a voyage into the past. He saw only one way forward—the microscopic route.

On a subatomic scale there are many options for time travel. But how could we use any such option to construct a usable time machine? Human beings are too large by many orders of magnitude to take advantage of these small gaps through the time barrier.

It was the concept of the arrow of time that paved the way towards the answer. Since Newton, most physicists have believed that the arrow represents a flow of events from past into the future. If you smash a cup, it breaks into pieces. The pieces do not assemble back into an intact cup again. Indeed, the whole concept of entropy, the state of disorder in any system, has become established as a rule of nature from such a logical conclusion.

At absolute zero, for example, there is no molecular motion and the entropy of a system is zero. As things heat up, motion increases and molecules fly in random order, increasing the state of entropy. Ice has molecules locked up with a little motion and just modest entropy. Water is freer to flow around, has a greater state of disorder, and hence has increased entropy. Steam has molecules that move far and wide and could spread across a very large area as the gas dissipates. Its entropy is even greater. As heat rises, so rises the level of entropy.

While the concept makes sense, the more we discovered about how the universe works, the more we found that the laws of nature did not seem to require it. Reverse the arrow of time and these laws of nature work just as well. So could the arrow of time flow in the other direction for some parts, or other versions, of the universe? Are we merely perceiving the flow of time from past to future in our local space, or is it just a mental convenience? Perhaps time could flow both ways. Of course, the notion that the arrow of time could be reversed—with time moving from the future into the past—has great implications for time travel.[79]

Until the twenty-first century this strange idea was deemed very unlikely. Entropy decrees that any universe where time flows in reverse would need to see matter re-form out of chaos—the shattered cup would re-create from its many pieces—though any slight variation would wreck such a reassembly process. Creating chaos from order is much easier, because there are many paths that debris can take when it splits apart. Re-forming order out of this chaos is far more difficult because there is only one way in which all the elements can recombine to be perfectly aligned. A bull can wreck a china shop in a thousand different ways. But it is still wrecked. The china shop would take billions of years to put itself back into the original order through any random motions of its atoms, because they all must follow one precise track to rebuild the original store. Only the intervention of a controlling force to order things— such as a shopkeeper sweeping up to a predetermined pattern—will achieve a task that nature alone would find all but impossible to accomplish.

But Lawrence Schulman of Clarkson University in Potsdam, New York, saw a way out of this apparently intractable dilemma, and inevitably it involved time travel. It also involves the fate of the universe.[80] Most scientists accept that the universe has expanded from a single super atom that erupted as a big bang billions of years ago. It is still expanding—as far as we can measure—with all galaxies rushing away from one another. Science believes that it will reach a point far into the future where the galaxies will start to contract again towards what might be called a big crunch. When contraction starts, the arrow of time will reverse and events that today seem to us to go from past into the future would appear to head backwards through time. However, anyone living in a contracting universe would not notice anything odd. Events would happen normally since their arrow of time—although reversed for us—would be consistent throughout their lives.

Remarkably, Schulman argued that the parts of the universe where time has reversed could already be all around us today, and some of them may be relatively nearby. In fact, dark matter fits the pattern of what to expect from such reversals, because it has gravitational effects that we can measure. This is a key reason why we believe dark matter is real, even though we cannot see it or measure it directly. Schulman suggests that this matter comes from the far future of the universe when the reversal of the arrow of time has already happened and all matter at that point is moving backwards through time towards a big crunch. During such contraction the stars will have burnt out as dark cinders—meaning that we could not see light from regions of reverse time, just feel their gravity. They would be quite literally "dark matter."

Another possibility, Schulman suggests, is that dark matter might result from collisions between normal matter in our familiar expanding universe, where time's arrow flows from past to future, and matter that is running backwards from the future towards us. This massive temporal collision would create regions of space where time is effectively eliminated, a sort of "null time" void. It would be like two rivers flowing towards each other with equal currents, meeting head-on and effectively standing still. Dark matter seems to have many of the properties predicted for such voids, making them places that all would-be chrononauts might wish to explore.

Needless to say, Schulman's research created much controversy when it was published. Raymond Laflamme at the National Lab in Los Alamos, New Mexico, simply could not accept that any universe could reverse back into a state of perfect order without a guiding force such as a cosmic shopkeeper to sweep all particles into the correct alignments. Others, of course, remind us that such a shopkeeper may exist—God may be minding the store![81]

Quantum mechanics may prove to be the crucial factor needed to solve this conundrum. Look at subatomic space, says physicist Murray Gell-Mann, where things are so different before and after the wave function collapses and a single reality comes to be the one that we experience. These subatomic events seem to happen in a "time neutral" state where past and future lose all meaning and what we see as events from the future can directly affect events from the past.[82]

Recall the double slit experiment conducted in Munich. The result seemed to depend upon decisions that the experimenter took after the experiment ended. How on Earth can you change the past by a decision taken in the future? Only if time indeed reverses.

At one point, faster-than-light particles—or tachyons—were thought to be involved in the process, beaming data from the future into the past. But nobody has provided solid proof that tachyons exist. Now Schulman and Gell-Mann were demonstrating that on a cosmic scale information from the future might be routinely traveling backwards through time to interact on a regular basis with its own past self (which, of course, exists in what to us is the present). But how does this time travel occur?

A breakthrough made in 2001 by Huw Price at the University of Sydney took us one step closer to microscopic time travel. His concept involves what we might call handshakes from the future. He starts with the premise that the arrow of time is at least partly an illusion formed by how we view reality. We extrapolate from what we see and assume that everything in the universe flows in the same direction—from the past to the future, but this is not an assumption that quantum physics bears out. Indeed, to interpret some of the most puzzling experiments, we have to accept that signals from what we call the future are capable of affecting what we call the past. The error we make is understandable, as we judge reality by what we see, not by what we can never see.

The problem of information beaming backwards through time is not a modern problem. When Maxwell developed the equations that proved how electricity and magnetism were linked in the latter half of the nineteenth century, he came up with some puzzling results. One set of answers revealed how a wave of energy moving out from the electrons at the speed of light conveyed electric charge. They went, as might be expected, forward through time. However, the second set of results from the equations produced bizarre energy waves that seemed to come from the future, moving backwards through time and converging with the electrons.

In the 1880s this made absolutely no sense, as Einstein was still decades away from proving relativity. We could only interpret the arrow of time as having a single direction. It was impossible for us to get our heads around a reality where time traveled in a direction opposed to the one we experience throughout our lives.

Most physicists, including Maxwell, assumed it was just a mathematical quirk and that time traveling energy waves had no reality. The idea wasn't taken seriously for nearly a century, when American physicist Richard Feynman started using these long forgotten quirks of the math to calculate how antimatter particles might behave. These could also travel backwards through time. Only when you produced a complete picture diagram of how particles and antiparticles behaved within subatomic space was it possible to make any real sense out of how matter interacted on a tiny scale.

Price revitalized this idea by trying to explain the results of the still troublesome double slit experiment as seeming to involve data being sent back through time because, quite literally, that is what happens.[83] At a quantum level, all energy waves have two parts. One is known as the "offer wave" emitted by a particle, a photon in the case of light being passed through the double slits. This travels into the future in the usual way, following the familiar arrow of time. However, it always interacts with a second wave coming backwards through time from the future to the past. This is called the "echo wave," since it echoes the signature of the first wave but in a reverse direction. From its perspective everything would seem normal, as it exists in a space where the arrow of time is the reverse of ours. To us this echo wave is time traveling backwards. The interaction between the waves determines the reality we see in the present. This interaction is known as the "handshake" between these two waves.

In the double slit experiment the offer wave beamed out by the photons must go through both the slits because it is possible that either or both might be open. All possibilities have to be explored at a quantum level, so both these paths are followed when heading forward via the usual arrow of time. However, if we make a choice and allow the photon to pass through one slit, the echo wave coming back from the future will only follow the path through that slit since from its perspective it only needs to pass through this slit because the other one is shut. The handshake between the echo wave and the offer wave reinforces the signal for just one slit because both waves pass through it. But only one wave (the offer wave) goes through both. The single slit reality is emphasized by the echo wave and crystallizes out.

Some physicists were unsettled by these ideas, not the least because the dreaded paradox seemed to rear its head again. It is almost like saying that an event will occur because an event has occurred, leaving cause and effect shrouded in confusion and chaos.

This view of how quantum particles interact argues that reality is what we experience because of a choice we made. Choose another path and reality will alter. And this is just as true if the choice we make comes *after* the outcome. Actions we take in the future must be able to alter the past.

Such an idea seems absurd, scientifically and otherwise. We all must regret errors we made in the past, but we know that we cannot undo them just because we wish we had acted differently. Surely no amount of decision taking after the fact will actually cause them to go differently in the past. We are so conditioned to think in terms of the past as an unchangeable entity that anything else seems absurd. Yet, time travel into the past would be fraught with such absurdities.

Price has a way around these objections and it will sound either trite or inspired, depending on whether you are willing to look beyond all that you have grown up believing to be true. He says that we interact with the past all the time; it's just that we do not notice the changes made in past events. Why? Because to us these events are part of a memory and from our present-day perspective we have long lived with them in a form that incorporates the changes that result from our future acts. In other words, we do change the past by what we do right now and the past is constantly adjusting to take account of these changes. However, because the past is the past, our consciousness will remain unaware that we have brought about such changes because whatever we do only serves to re-create the past in the form that our memory tells us we have experienced it.

If you are confused by all of this, so are many scientists. We simply lack the mental agility to think readily in such time traveling terms. Needless to say, we may have to gain this skill if we are ever to use such knowledge in a practical time machine.

Can Price's research help us achieve time travel? Perhaps, because it also reveals the secret behind quantum entanglement. This mysterious process pioneered by Einstein seven decades ago features twin particles that streak apart far across the cosmos yet seem able to communicate through time and space. They are actually time traveling, according to Price. They send echo waves back from the future to interact with their offer waves heading towards that future. What results is an interaction free of time and space.

Quantum entanglement turns out to be the ignition switch that turns on a time machine.

2002
TWISTING TIME

Most researchers trying to build a time machine are motivated by the sheer desire to make it happen. Ron Mallett is different. This physicist has a very clear reason for wanting to achieve time travel, and it is deeply personal.

When he was just a lad of ten, almost fifty years ago, his father died suddenly. He was only thirty-three but had smoked profusely. Mallett felt cheated out of a relationship, but his fascination with science fiction stories offered him hope of an escape route—time travel. If he could build a time machine, he could travel back to before his father's death and make him quit before cigarettes claimed his life.

This sounds like the plot for a blockbuster film and indeed a similar scenario was featured in the movie *Frequency*, where a ham-radio operator makes contact across the years with his dead father. Thanks to this temporal two-way link, he not only gets to talk to the man that he lost too young but also can try to warn him of the tragedy that killed him. Mallett's desire to save his father predates any movie, however.

Mallett became a professor at the University of Connecticut and began a series of experiments and theoretical designs that aimed to make travel into the past a reality.[84]

Once he began working through the math, Mallett became satisfied that the paradoxes of time travel were not insoluble. Paradoxes only become a difficulty if they lead to historical absurdities. Avoid that and there is no problem.

If you travel back to the day that you first constructed your time machine, then smash it to pieces, you invoke a paradox. Otherwise your destruction of the time machine prevents you from using it to reach the past where you then destroy it. But it is a paradox only if the laws of the universe let you time travel and carry out this destruction. They may allow travel but not a change of history.

Every action you take in the past may simply work to avoid the destruction of the device. You set off towards it, trip over a carpet, break your leg, and end up in the hospital, not being released for several weeks until you are due to use the time machine to return into the past. This prevents you from doing anything that would upset the laws of time. We assume that free will must give us the ability to bring the paradox about, but perhaps the very idea of a paradox is an illusion because hidden rules, while not stifling our ability to make a choice, limit the nature of those choices and the form of their outcome.

Heartened by this thought, Mallett began to work on a machine that can warp space with the time-bending properties of a black hole. He was aware that Frank Tipler's efforts to build a time machine with an artificial black hole had faltered due to the enormous energy required and the attendant risks involved in its creation. So he sought some other force that might provide similar time-bending effects but would be easier to accomplish.

Anything that possesses energy, noted Mallett, has the capacity to affect space-time by warping its structure. Some are just more effective than others. To make a time machine, you need to strike a balance between effectiveness and the ease with which we can produce sufficient levels of this energy to create a time warp.

Mallett decided to use light to warp space; since light has no mass, it does not require huge amounts of energy. Indeed, the ongoing experiments with slow gas using condensates had proven that light could be manipulated to such a degree that its progress could be completely stopped, effectively freezing time. Mallett saw this opportunity and has tried to build on it to develop time travel.

After first experimenting with a set of lasers to create a vortex in the lab, Mallett settled on a plan to create a time machine using a set of powerful lasers placed in orbit. Powerful lasers have enormous coherent energy and if rotated at great speed in opposing directions, can distort the gravity in their vicinity. By using a very cold bath of atoms close to absolute zero, you could restrict the speed of the laser light to very low levels and increase its intensity at the same time.

In theory, Mallett's prototype time machine should succeed in warping space-time and create distortions in time within the vicinity of the spinning lasers, though they would be very trivial distortions initially. But the lasers have to rotate very quickly if they are to cause gravity to warp by any significant amount. That is the biggest hurdle Mallett faces, because producing such speeds requires ever-greater amounts of energy.

Mallett believes that with sufficient energy, you can force time and space to effectively swap places inside the zone affected by the lasers. Here time may possess a spatial dimension with up-down, left-right coordinates. The switchover manufactures a sort of timescape—time with its own landscape. And within this timescape, you should be able to move spatially and travel through time. Suitably rotating beams should make journeys into the past quite feasible.

One extraordinary consequence of such a device is that you could walk into the ring vortex of warped space surrounding it and then pass yourself on the way out. Because the "you" heading out will be from the future having traveled back in time, it will meet the " you" just walking in.

Which would be the real you, and what would happen if you chose to shake hands like subatomic particles during double slit experiments? The experiment poses countless weird questions, but those handshakes through time between subatomic particles seem real enough. There is every reason to suppose that humans could do this also, if the energy problem can be solved.

Mallett started to build his machine in 2002, sure that he could send particles through time to visit their own past—although possibly at first by just fractions of a second. Bigger time jumps are possible with this equipment. It is simply a matter of the technology providing the necessary increase in power. With breakthroughs occurring with increasing frequency, by the time Mallett's device is up and running, it might be possible to demonstrate time travel of subatomic particles by seconds, minutes, or even hours. An advanced version of Mallett's device might allow the particles to be sent into the path of the rotating lasers almost like riders sliding down a fairground slide helter-skelter. Mallett sees no reason why this time machine should not carry humans, but in initial experiments only relatively low-mass subatomic particles are likely to be tested because the energy input required is then vastly reduced.

On the problem of temporal paradoxes, Mallett sees two possible ways out. Either there are multiple realities, allowing us to change events in the past that we visit without altering the future in the parallel, near identical, "reality" from which we have left. Or undiscovered forces within the universe might exist to prevent paradox-creating events from ever coming to fruition.

Intriguingly, Mallett thinks that once he has constructed his machine and turns the key, the floodgates will open to an invasion out of time. Particles may well suddenly start to appear as if by magic and coming from nowhere. These will be products of experiments he has yet to carry out. Someone using the device next year, or 20 years from now, will be producing the particles that are coming back through time to shake hands with themselves. Even though time travel is only possible back to a point when the machine is operational, the moment it is turned on opens up the possibility of visitation from any point in the future.

Once Mallett gets the machine up and running, we are likely to learn many things about the nature of time travel that at present are nothing more than science fiction or conjecture. We are entering an extraordinary period in the history of science. The moment time travel occurs, everything we consider normal will change.

If Mallett discovers how to control the flow of particles, he might then be able to convey meaningful information through the time barrier. He then might even get a message from the future when he switches on his machine. Will the future Ron Mallett be able to advise today's Ron Mallett on the best ways to set up his equipment? Unfortunately, even if Mallett succeeds in creating a version of this machine that will allow travel into the past, it is not likely to fulfill his wish to see his father again. Since, as far as we know, there were not any working time machines in existence all those years ago, any real hope of traveling back to that era is precluded. Unless something unexpected happens when Mallett makes his first successful experiment, his dream may prove to be a dream too far, even for a man with a time machine.

2003

BEAM ME UP

If you have watched any of the famous science fiction TV episodes of the *Star Trek* franchise, then you will know that the characters hardly ever bother with tiresome flights down to a planet when their ship is safely ensconced in orbit. Instead they make this short hop using a *transporter* beam. After the latest nasty alien has been routed, the crew members say "beam me up" and promptly dematerialize, only to reappear in a trice back onboard their spaceship. They have traversed large distances in the blink of an eye.

Yet if they make this journey instantly, then they must travel through time. In *Star Trek* fiction, the transporter beam can only be used to cross relatively modest distances (a few thousand miles) and so any time-jump that results is much too small to be noticed. But physics dictates that a time-jump must happen nonetheless. And if the journey through space is longer, so is the time trek.

Amazingly, this is an area of science fiction where real science is quickly catching up. Remarkably, the first primitive teleportation devices have already been tested, using data rather than objects for the moment. In many respects the process is not unlike sending a fax—the principle behind which science has known about for a hundred years. In a fax, the information that you wish to transfer is broken into digital bits of data, which are then sent at very rapid speed directly to another fax machine. Once the information is at its destination, another fax machine re-creates a copy (or facsimile) of what was beamed down the wires or, these days, perhaps digitally straight through the atmosphere. The word *copy* is really the key here.

It is also the reason why there has not been a rush of human volunteers to be guinea pigs to be beamed across the universe. Living things will not be so easy to package up and send as information. Even if there is no technical barrier to teleporting any *object*—including a human body— the human mind may prove a little trickier to transport than physical matter of the body and brain.

In 1998 researchers at the California Institute of Technology succeeded in breaking down and then sending a beam of light a few feet across the lab. This was the world's first successful teleport, and it worked by virtue of the fact that light is an energy emission and, like other EM radiation, can be transported and reconstituted into its original form at the other end.[85]

Sending a fax resembles the transmission of a sound wave created by a concert musician in a distant theater. If this is transformed into another kind of EM energy (radio waves) and beamed through space, it can readily be turned back into a semblance of the original sound by your radio set. You hear that sound as if it were "live" and in its original form. Of course, we know that the sound coming out of a radio is just a *re-creation* (good or not so good, according to the quality of your receiving equipment). Nobody believes that the receiver in your living room replays the exact sound wave that emerged from that distant piano. But we accept it as if this were the case.

Light can be broken down in much the same way, then beamed to a receiver and re-created as light that mimics the original. But the crucial thing to remember is that this new copy of the light ray will involve completely different atoms. The light beam that is re-created is no more the original than the piece of paper emerging from your fax machine is the same piece of paper put in by whoever sent this message to you from somewhere else. They *look* alike because the *information* that forms the message is re-created in identical form. But they do not have a single piece of matter in common— no atoms or subatomic particles are physically beamed through space by sending a fax message. And that's what happens when you teleport an object: The receiving machine creates a copy of the original; it does not rebuild it from the atoms of the thing being transmitted.

Physicist Jeff Kimble at Caltech believes that the principle his team used in their prototype experiments could be applied to send anything and that tests beaming primitive living organisms, such as bacteria, might be possible in the first decade of the twenty-first century. Two labs in Europe used the same technique to prove that other forms of EM radiation could be teleported as well. Ignacio Cirac at Innsbruck University in Austria headed one successful replication. In September 2001, he spoke of a coming revolution in the transfer of information that would relegate both the computer and the fax machines to the museum. The new method—a quantum communication system—could instantly send things from one place to another simply by encoding and decoding data streams of EM radiation.[86]

Let's say you order a new toaster. In the old days, you had to phone up the store, ask them to send it, and wait for it to arrive at your door by delivery truck. Today you can access the firm's website on the Internet, but once this instant method of ordering is done you will still have to wait until the toaster is physically transported to your address.

This will not be true when you have a "quantum teleport" in your living room. Instant ordering will be followed by instant delivery. The toaster company 1,000 miles away will transmit the toaster's data and a toaster will "appear" on the mat of your quantum teleport device. It will form from out of the matter available to the machine—your teleport device will use the information template beamed to it as the blueprint that assembles the product. We are, of course, many years away from that level of technology, but the science for it is largely now in place.

By 2002 teleportation had taken the next great leap forward when the advantages of using quantum entanglement were perfected. It was quickly evident that this provides the best method of data transfer. The largest data stream so far teleported was sent with this process by physicists at the University of Aarhus, Denmark. Eugene Polzik explained what they had done: "We have produced entanglement at a distance. Then we can untangle the two objects." This is likely to be how the more complex teleport machines will work.

What did these scientists do? Entanglement occurs at a quantum level when you alter the properties of one particle (e.g., a photon in a stream of light) and its twin particle "magically" responds. What you do at one location causes an entangled particle at another location to adapt in unison, irrespective of where that other particle is positioned. The transfer of data occurs instantly, even faster than light.

The American Institute of Physics described the first successful use of quantum entanglement to teleport sample particles from within a cesium gas cloud that was then split apart. By firing a light beam through one half of the cloud, the physicists altered the properties of the other half. The result proved that even across the smallest measurable time scale (one millisecond), the two clouds altered their properties immediately and together. When the first cloud adapted as the light passed through, the second changed in compensation at precisely the same moment. When the second changed as the light reached it, the first then adapted back to match.

As the theory of "quantum action at a distance" shows, particles in both clouds were linked together via those hidden dimensions of time and space embedded within quantum reality. The particles in cloud one were not beaming out a signal to those in cloud two, like an unseen radio wave. The compensating change occurs because at a quantum level all these particles are somehow ineffably linked through time and space in ways that we are unable to detect.[87]

For the first time in history, significant amounts of data were transferred through space without any visible physical medium allowing it to happen and completely free of time restraints. If we can now tailor the changes to act "intelligently" and send a required message, then we can initiate specific changes and transmit complex data through space.

A more refined experiment at the Australian National University in Canberra in June 2002 extended the research and transferred many photons of light at once instantly through space. Ping Koy Lam, who masterminded this test, was upbeat about its potential uses for computing. By sending data this way, he believes we will boost the ability of computers by many orders of magnitude in the coming decade, making possible huge tasks that are too onerous for our computers today. But he was less convinced that we would ever send living beings through space in a teleport machine.[88]

Quantum entanglement transmits *information*, not things. It adapts distant particles so that they can mimic the events that you trigger. The "copy" of the data formed at the distant location is a re-creation brought about through the entangled link through hidden dimensions across space-time. Therefore, you could not actually send a Ming vase through a teleport device, but could attune entangled particles in distant atoms to re-create an identical vase wherever you transmit. As long as the vase looks the same, it seems to be a successful transportation.

But who is going to volunteer as a living person to be rebuilt in this way? What happens to the "identity" or soul of the person who is supposedly being transmitted through space? Is the new copy created at the receiving station another "you" or just an empty soulless shell? By deconstructing the atoms at the sending station, does that person who is disassembled die to then be reborn at the other end in the form of a zombie?

Nobody will know the answers to such questions until such an experiment is actually performed. Nevertheless, we do have some evidence for the continuity of identity even when the atoms in a copy are not the same as those in the original. Every single human being living into adulthood consists of completely different atoms from those with which they were born. That is because we lose countless atoms every year as our body sheds skin, deposits waste matter, and has organs break down. The identity of the self is maintained through what amounts to a full body cell re-creation process throughout our lives, presumably because something imprints a pattern onto the new atoms—no doubt through EM fields channeled within the brain. It is not inconceivable that a teleported person could remain intact even if they are beamed into brand new atoms even as the transmission process shreds their old ones.

This new technique has many other possible ramifications. For example, by using quantum entanglement a modified copy of an organ or body part could be made without any of the genetic defects of the original. Future medical procedures might use this method to initiate cures for seemingly incurable diseases. But, given the huge ethical questions over cloning and stem cell research, it remains to be seen whether such an experiment would ever get moral sanction.

And what does this have to do with time travel, you wonder? Long-range teleportation will involve serious time travel, because quantum entanglement uses time traveling particles. If you transfer from Earth to Mars, for instance, you will instantly arrive at your destination, but the light conveying the image of your departure will take several minutes to cross the void and catch up. Recall the axiom—beat light in a race and you time travel. In this case you go back into the past by whatever gap is introduced by the speed with which light makes the same journey. It may be minutes. It may be longer. The farther you teleport through space, the more you time travel.

Science fiction has already latched on to the potential of this state-of-the-art research. Michael Crichton's 2000 novel *Timeline*, made into a movie in 2003, took note of the earliest of these teleportation experiments. The plot involves a time machine that deconstructs people by converting them into energy and projecting that energy through the hidden dimensions within microscopic space-time.[89] Crichton's fictional time machine is basically a teleportation device.

And so it may prove in real life. We could one day be sending basic organisms such as microbes through time using teleportation, and may be able to send higher life forms a few minutes into the past before the decade is out.

Real teleportation is real time travel. The day of the time machine is about to dawn.

2004 CHRONONAUTS

The United States won the space race against the Soviet Union when it successfully landed humans on the moon in July 1969, fulfilling President John F. Kennedy's promise of almost a decade earlier. The Soviet Union is now a part of history but a new battle is waging to take humans through the time barrier. This race involves experimenters, professional and amateur, from every corner of the planet, but it is likely that practical time travel machines will be developed first in a nation that can afford the costs involved in such a breakthrough. As such, the real fight involves research teams from the United States, Japan, and European Union scientific institutes—although significant contributions are coming from outside sources as well. It is still an open question as to who will win this extraordinary race.

Although many of the most enterprising time machine designs come from the United States, making this country the odds-on favorite, shocking news coming out of Russia suggests that it now may be ahead of the pack. One Russian scientist even claims to have constructed and tested a machine and, he insists, successfully broken the time barrier—all using *human* pilots. If true, then the first chrononauts have already made this momentous journey. The race may be over, although not before a lengthy review of the data by science.

The man making this remarkable claim is Dr. Vadim Chernobrov, who works at the Moscow Aviation Institute. He became interested in time manipulation, as he termed it, while working on his doctoral thesis to create a flying machine with a revolutionary propulsion system. The effects on the flow of time were side effects of his calculations and these soon became the primary focus of his research.

Unlike many Western scientists studying time machine technology, Chernobrov appears to have a support structure. Time had been studied as a branch of physics in the Soviet Union since the 1970s because the space program there had a different focus from that of the United States. Because Russian space crews made it a point to spend months at a time in orbiting space stations, they are very familiar with time dilation effects. These Russian cosmonauts have time traveled into the future by several fractions of a second. This may seem tiny, but nobody knew what effect this would have on the human body, and the need to know ensured that research grants studying the nature of time through direct experimentation were much more readily obtainable in Russia than in the West. In addition, Russian science has been willing to consider subjects, like time travel, that Western science has, in general, tended to ignore.[90]

Since the fall of communism many of these weird programs appear to have been terminated. However, one of the construction centers used in the research, an aerospace plant at Khrunichev, has teamed up with the United States defense authorities in the new era of friendship to work on various secret projects. This is the genesis Chernobrov claims for his work. He also had access to secret Soviet records dating back to the early days of the Cold War. From these he says that Soviet scientists had come to believe that during World War II, Albert Einstein had been part of United States military experiments that accidentally led to time travel, even though their intention had been to use magnetic fields to try to warp radar signals. This research is considered pure folklore in the United States.

These files supposedly persuaded Chernobrov that Einstein had destroyed all record of his research, because it had enormous potential for use as a super weapon. That research had as its starting point Tesla's 1895 experiment. The nature of space and time means that even modest explosions could be devastating when channeled through the hidden dimensions enfolded into microscopic space. Einstein did not want to be remembered for allowing a time travel weapon to be built, adding to his major assist in creating an atom bomb. But the Soviets were happy to support further research, although Chernobrov has been able to warn about the risks only since the fall of the USSR brought him freedom to publish and question.

The time machine that Chernobrov has designed is modeled on the famous Russian doll, a very old toy that comprises a number of figures in traditional costume, each built slightly smaller so as to slot inside one another. For his prototype, activated for the first time on April 8, 1988, Chernobrov built a series of capsules that fitted inside one another, with four or five proving to be the optimum number. They were molded into spheres with a maximum diameter of about three feet; the surface of each sphere had flattened edges causing it to resemble a multisided die. They were constructed out of electromagnets, being operated in a superconductor mode at extremely low temperatures. These temperatures were barely above absolute zero when Chernobrov first started. Even today only modest increases in operating temperatures have been made.

Chernobrov has constantly modified the dimensions of his spheres, the materials used, and the size of the EM fields that circulate within them. The innermost sphere is the "payload capsule" into which anything to be sent through time is located. It is only a few inches in diameter. This receives the maximum bath of EM energy when the device is activated, although there are effects to a decreasing degree in the outer spheres and even residual effects in the room surrounding the device.

In principle, the machine's interacting EM fields, and their carefully designed twisted elliptical shape, distort space-time in a way that allows travel into the past and future. The extent of this "time manipulation" varies, but it would require huge energy (hence the need for better superconductors) if travel through days or years is to be achieved. Even so, the intrepid researcher claims to have progressed from barely measurable results to ones that qualify as genuine time travel.

Initially Chernobrov simply used measuring devices, such as atomic clocks, to check how much time passed within the inner sphere compared with the amount ticking by in the room outside the machine. After finding a combination of sphere dimensions and field strengths that worked best, he managed to record a time difference of half a second per hour inside the payload capsule.

In other words, for every 24 hours that went by in the lab, only 23 hours, 59 minutes, and 48 seconds ticked by inside the sphere. This was genuine time travel, greater than ever reported in other experiments, but still practically insignificant. By the early 1990s, however, he had managed to increase the difference considerably, so that 23 hours, 59 minutes, and 24 seconds were now passing in the sphere for every day in the world beyond. At this stage Chernobrov decided to risk sending the first living creature through time.

Given the modest size of the payload capsule, only small creatures could be transported. So, by default, insects and ultimately mice were the world's first chrononauts. Unfortunately, they did not survive the trip, despite the very small time shift. Indeed, even some of the lab technicians standing close to the device during the experiments suffered nausea, dizziness, and mild skin blisters. Only after Chernobrov made extensive modifications to his equipment could mice time travel without apparent ill effect.

By 1996, when Chernobrov presented his results at a science symposium in St. Petersburg, he had been able to use the device to travel both forward and backwards in time. Time travel into the future proved to be more successful than time travel into the past.[91] Slowing time is akin to travel into the past. However, if you sit in a pod for two days and then emerge to find that a few seconds less time has passed for other people than for you, it would be difficult to persuade yourself that you had really traveled back in time. Travel into the future is more obviously real, because if you spend just a day in the device and perhaps a week has passed in the world outside then you would certainly believe a significant time travel journey had occurred.

Chernobrov is talking about differences of only seconds at the moment, certainly not anything as long as a week, although he is sure that this is only a question of more energy and further engineering. Of course, it requires even more energy to increase the size of the payload capsule to allow a brave human passenger, but there is no theoretical reason to prevent this from happening.

Chernobrov's research has concentrated upon trying to discover the laws of time travel from his numerous experimental runs. He has shown that outside factors can often create changes in the level of time distortion. Some of these have proved quite enlightening.

The presence of electrical fields in the atmosphere seemed to be one of the most significant variables. Electrical storms cause greater levels of time distortion. There also seemed to be a link with gravity and tidal forces varying with the phases of the moon. None of this is very surprising given what science has discovered about the nature of space and time in recent years. And it may cause us to think twice about homemade time machines such as Bassett's lightning-in-a-bottle-fuelled bio-energizer.

The most remarkable experiment in Chernobrov's several hundred test runs produced a time shift of 12 minutes in a 24-hour day—much greater than was regularly achieved during other tests, but he has not been able to find out why that particular experiment was so unexpectedly successful.

These tests have led Chernobrov to view time as if it were a tree. The moment we call "now" sits where the thick trunk sprouts all of its branches. The past is the tree trunk and has just one unified course climbing upwards. The future, like the branches, unravels higher still in various directions. If you go into the past and then return to "now," the now that you revisit will be different from the one you left. This view conforms to the many-worlds theory of quantum physics.

In 2003 Chernobrov took this already extraordinary story to new heights when he claimed that he had completed the ultimate challenge, making the first tests with a human chrononaut. As so little is yet known about the potential risks involved in time traveling, the first volunteers stayed inside the inner capsule for no longer than 30 minutes to minimize the potential side effects on their bodies. But all reported feeling some weird sensations, rather like being taken out of time and encountering space as "enfolded." They had difficulty finding the words to adequately describe these feelings, but they included a sense of being both "here" and "not here" at the same time, a sense of dissociation that may prove to be the price you pay for traversing the highways of time. In fact, the experience was much like Tesla's in 1895, when he described how he accidentally stepped out of normal space and time and entered a seemingly new dimension where time possessed space-like features.

Chernobrov's work has yet to be independently confirmed let alone replicated, but if we take him at his word, then one thing is clear. Time travel into the past and the future is no longer a question of "what if?"

The bid to break the time barrier was launched by Herbert George Wells more than a century ago, setting mankind on a journey into the scientific unknown. Now, 110 years later, Wells's startling question may finally have been answered and, as we push forward into a future that few could have imagined, the appropriate question that presently confronts us is—"What now?"

BEYOND
TIME'S LAST FRONTIER

Despite the claims of Chernobrov, the search for time travel goes on. In particular, we still await the first practical, reusable time travel device that involves levels of time displacement greater than a few minutes. The world is unlikely to accept that the time barrier has been smashed until a scientist reveals a machine that produces significant time-jumps and which anyone can use. Just who will earn that place in scientific immortality remains an open question, but many of the names that have figured in this book have paved the way for a day that now seems increasingly imminent.

Yet it seems improbable that we will all have time machines in our garages any time soon. Time travel is going to be very difficult, dangerous, and costly. Only a government-funded project making a concerted effort is likely to make it successful in the near future. Electricity is difficult and expensive to generate in large, usable quantities. But it exists in abundance amidst the 1900 thunderstorms that rage simultaneously over the Earth at any given moment on an average day. Tap that power, as scientists like Tesla have sought to do, and all the energy you would ever need would be available to you.

Of course, our task would be greatly facilitated if natural time travel already exists. Do wormholes form where humans might be able to make use of them? Perhaps the way to break the time barrier is right in front of our noses. But how would we look for any wormholes large enough to be useful as a time machine, and why would they not already have been recognized by science? When scientists are not able to directly observe a phenomenon, they look for something that is likely to follow as a consequence of the phenomenon's existence. That's how black holes were found. This may also be the way to find naturally occurring microscopic wormholes that might exist in our own backyard.

But what would be the consequences of small wormholes operating sporadically within the Earth's atmosphere? Nobody knows, but we can make some reasonable assumptions. There would be gravity anomalies, space-time distortions, and evidence for spontaneous time slips and spatial relocations. Other side effects might include light rays behaving in a very unexpected fashion when in close proximity to such an open wormhole.

The fact that physicists have not accumulated strong evidence for such phenomena seems to suggest that they do not exist. But the phenomena would likely occur only sporadically, and if witnessed, would probably attract the sorts of paranormal interpretations that would send most scientists running in the other direction. As a rule, scientists are hostile to such subjects and to an extent their caution is justified. But it is wise to remember that many natural phenomena that science has come to understand—such as weather-induced mirages or ball lightning—were reported for hundreds if not thousands of years before we had a scientific explanation for them. For centuries they were widely considered to be supernatural in origin by anyone who chanced upon them. The fact is that weird stories often contain a kernel of truth waiting for our knowledge to catch up (see the appendix for more on this possibility).

But all things considered, the road to time travel sooner rather than later probably lies almost where we started. Frank Tipler at Tulane has spent thirty years trying to find a way to safely make an artificial black hole time machine. So far he has failed. But in the meantime technology has progressed in leaps and bounds. A growing number of time travel physicists wonder if the extraordinary explosion in computer processing power will soon fulfill the dream that H. G. Wells had back in 1895 when he attempted to create a time travel simulator. However, a computer-based time travel simulator would be incredibly more persuasive than Wells envisaged. So persuasive, in fact, that it might as well be real.

Over the past thirty years, breakthroughs in processing power and data storage have turned what were once slow, lumbering, and gigantic devices into rapid, miniaturized systems that can fit into almost anything. But changes lie ahead for computers that will revolutionize the devices and what we can do with them.

We are beginning to use light to store vast amounts of data. This is one reason the experiments to slow light and freeze data streams are so exciting. It may seem a little odd that scientists often talk in terms of computer applications, rather than time travel possibilities, when explaining why they run these experiments. But that is because computers of the future will make today's computers look like stone tablets. Huge profits await the companies that pioneer this technology. A computer that works a million times faster or stores a billion times more information will make countless technological advances possible.

Time travel will be one of those possibilities, and its implications are staggering.

Anyone who has played a virtual reality game on computer will know how the simulation of an environment has become so sophisticated that computers are now used to train pilots, surgeons, and other professionals in their jobs. Once inside a virtual world created by the computer, trainees find real working conditions that allow them to practice and make mistakes that will not prove disastrous. Computer-generated environments can now match reality so perfectly that the illusion can be hard to shake.

We see much the same effect when it comes to computer-generated movies. The cheesy animations, landscapes, and characters of yesteryear can now be so realistic we may soon find it impossible to tell where reality ends and illusion begins. The next step will no doubt be to computer-generate long-dead movie stars and put them into new adventures, with undoubtedly contentious effect.

Time travel physicists now await the arrival of a super-computer that can create a simulation of the past so exact that it will be much like taking a real journey into that era. Computer characters could be programmed to act independently and interact with one another to "relive" past worlds and past events in front of observers. Indeed, it is not difficult to imagine that these simulated time travel experiences could be made so believable that we would be able to incorporate new generations of virtual reality equipment and let the observer become a participant in the events.

We can already link many of our senses to computer imagery in order to feel, smell, hear, and experience an illusion of an artificial reality. Future computers might be able to simulate time travel so completely that it will be as if the participant had actually stepped into a time machine and returned to the year of his or her choosing.

The virtual time traveler would then be able to take a safari among incredibly realistic dinosaurs, or attend the Coliseum in ancient Rome. Such technology would have hardly any boundaries. Several railroad enthusiasts who gather outside my house to watch occasional steam trains pass by say that they would give a year's salary to be inside a believable computer simulation of a busy station in the 1950s, during the heyday of their beloved mode of transport. Doubtless others would put this technology to very different uses.

Some people might regard the notion of becoming involved in computer-generated environments as something of a cheat, since it does not involve real travel into the past. But it follows in the great tradition of Wells and his immediate realization of how the public would relish such an opportunity. It would have none of the catastrophic consequences of getting too close to a rotating black hole. Nor would it stir up any prospect of reality-threatening paradoxes.

Perhaps more significantly, this kind of time travel is far more likely to be possible soon, given our present level of technology. It may, in fact, be only a decade away from commercial application.

But there is a sting in the tail with this type of time machine. If we simulate the past so perfectly and become immersed within an illusionary world, then virtual time travelers might believe that they are really living in the past. How can we ever distinguish between what is real and what is illusion if the gap between the two defeats our senses?

More worrying implications follow. What is to say that we are actually living in the twenty-first century? Might we not be part of some future simulation of the twenty-first century set up as a form of time travel experiment? This theme was, of course, explored to some degree in the science fiction movie *The Matrix*. But some scientists suspect that this idea may not be science fiction.

In fact, some computer programs have been run attempting to extrapolate on the current rate of progress of this technology. They reveal that by the year 2100 computers could reach such a level that a re-creation of an entire solar system would be possible in such minute detail that not even scientific instruments could detect a difference from the real solar system. Indeed, with a truly complete simulation, the distinction between the real and the imaginary loses the significance that it has for us today.

After all, everything that we know about how our senses perceive the universe says that we distil our reality out from the energy fields, quantum foam fluctuations, and statistical chaos at the heart of subatomic space. We apparently create reality in our minds as much as we simply observe it. We are architects of what we see, touch, and feel within the "real" world. If so, is it right to call a realistic, time travel-style, computer environment a cheat?

The race to break the time barrier has many twists and turns, and exactly which direction it will head next is impossible to say. It is a story that so far has no end. New discoveries and strange experiments are being reported every month. Further attempts to build a time machine will continue until one day reports of the first successful time travel will fill the media and the pages of the refereed scientific literature.

We must also face many critical questions about what we should choose to do with time travel if we can harness its potential. Science is not intrinsically evil. But, as biological or atomic weapons reveal, its discoveries can be used for evil purposes. The ability to manipulate time would provide a dictator with the ultimate doomsday device: allowing one to change the past or adapt the future until it suited his or her ends. We may find ourselves hoping that Hawking is right and there is some fundamental limiting force at work in the universe, if not to prevent time travel altogether, then to restrict its worst consequences.

Failing that, the scientific race to break the time barrier poses major political questions. It may be that as this research nears fruition supervision and vigilance by the intelligence services should follow. Although nobody would wish to see freedom of inquiry suppressed, a decision must be made on just how much freedom users of time machines will be allowed. Even something as simple as an automobile has had to spawn numerous laws to govern its sensible use owing to its potential for harm. But a single car cannot destroy the planet. Just one time machine in the wrong hands might be able to wreck the universe as we know it. These issues must be dealt with along with the more frivolous ideas about using a time machine as the ultimate tour bus or to rescue personal disasters in our past.

Human society will face many difficult questions when that first time machine is switched on. Like the first moon landing, the discovery of time travel will change our world. It will certainly be one of those defining moments in our civilization.

Appendix

DOES EVIDENCE
OF TIME TRAVEL
ALREADY EXIST?

This book is not about strange phenomena, but for thirty years I have been collecting cases of time anomalies from all over the world. I have done so because I believe that among the deceptions, misperceptions, and errors of judgment that make up any set of tales like these, there could be a handful of cases that represent the time travel equivalent of a rainbow before science came along to explain it. Time anomalies, like that rainbow, may be a clue towards new advances in real science.

Is there enough evidence for such time anomalies to suggest that humans do sometimes encounter something like naturally occurring wormholes? Surprisingly, the answer

Jenny Randles

appears to be yes. The evidence is much stronger than most scientists would probably imagine, especially as the data forms a very good fit with the physics unraveled during the race to break the time barrier.

These "time storms," as I call them because that describes their nature without presuming any origin, are almost always interpreted by the witness, popular media, and everybody else in misleading terms—often as alien abductions, ghosts, and various other paranormal phenomena. I prefer to strip these cases down to their bones, referring only to the hard facts that can be reasonably inferred from the evidence—minus the strange ideas that people tend to attach to them afterwards.

Here are just a few cases that illustrate why these events could be vital evidence for all those seeking to build a time machine. It is the pattern of evidence present in these cases that I find significant, more so than any individual case.

This first incident took place at 1:40 in the morning of August 27, 1979. Deputy sheriff Val Johnson was driving a two-year-old Ford down County Road 5, ten miles west of Stephen, Minnesota. Suddenly a brilliant glow appeared in the sky ahead, rushed towards the police car, and swamped it. Johnson reacted instinctively by applying the brakes, then everything went black. When he recovered, the light had gone, the Ford was 954 feet farther down the road (measured by an official accident inquiry), and Johnson immediately phoned his base office, where his conversation was recorded. The deputy sounded completely stunned. During the call, which took place at 2:21 a.m., he noted: "I don't know what the hell happened."[92]

The deputy was taken to the hospital, where he was examined at 4 a.m. The next day, he was driven to Grand Forks, North Dakota to see an eye specialist for a puzzling eye injury. His eyes were red, hurt when a light was shone into them, but healed remarkably quickly. The specialist felt that the huge ultraviolet light emission associated with the glow had caused the injury.

The case was exhaustively investigated, with scientific input from Allan Hendry, a skeptical researcher with the American UFO group CUFOS. Hendry flew to the scene when police contacted him for advice. Within 24 hours he had gathered all the evidence and interviewed the key witnesses. Weather data and radar records on low-flying aircraft failed to provide an explanation for the light. Ford Motors and Honeywell Inc. in Minneapolis examined the damage to the patrol car, one exploring the mechanical components, the other the electrical components. The damage to the windshield could only be explained by dramatic changes in the air pressure or gravity at the time of the incident. The air pressure or gravity inside the car must have increased and that outside the vehicle decreased. Nobody had any idea how this could happen, but it strikes me as rather similar to the Casimir Effect that seems to trigger antigravity and may facilitate time travel.

The other remarkable finding of the accident investigation involves time distortions. You might assume that the time delay between the deputy's experience of the event and the call to his office occurred because the deputy was temporarily rendered unconscious. However, both the electric clock inside the car and the wristwatch worn by the officer were working perfectly when the emergency services arrived, and continued to work perfectly during subsequent tests. But there was a problem. They were both running slow and by precisely the same amount of time—fourteen minutes. A study of Johnson's radio calls earlier that night showed no such time differences. Since the wristwatch was a cheap manual windup and the car clock an expensive electrical timer, what force could possibly cause them both to lose exactly the same amount of time? What phenomenon might actually cause such a range of odd events?[93]

This would be fascinating enough as a one-off case. But it is far from that. Many similar cases exist. Here is an incident that I investigated first-hand. The pattern is much the same. I am certain that the witnesses were unaware of the earlier case in Minnesota.

It occurred on October 8, 1981. Dawn Pendse, a Christian aid worker and wife of a British colonel, was driving through the Salen Forest on the Isle of Mull, located off the western coast of Scotland. With her were two American visitors, also Christian charity workers. One of them, Dwight, was driving and had slowed down to take photographs of the heather-clad hills. It was early afternoon on a beautiful fall day with clear skies when a dark mist suddenly enveloped their slow-moving car and seemed to solidify around them.

The car began to shake and a vibrating noise filled the air. A change in atmospheric pressure then created a suffocating heaviness inside the car that pushed the occupants down into their seats. Dwight recalls ducking behind the windshield instinctively. Dawn says that before everything went black she saw the mist literally suck all the light away, like it was drawing it out of the air.

Suddenly they found themselves on the road some distance away—unaware how they got there. The car was stopped and there was a total silence all around. The black mist had gone. But, just as with the Minnesota police car, there was evidence that it had been lifted off the road and dropped down again with force.

The sun's position in the sky and the fact that the stores were closing as they drove into town indicated that about three hours had passed. But they had not been asleep. It was as if they had traveled forward in time. Dawn's wristwatch was working but still showing the time as if it were three hours earlier—which it clearly was not. And the electric clock in the car and the quartz driven watches worn by both Americans had all stopped at the wrong time—three hours earlier. The quartz watches both needed new batteries and the jeweler who fitted them the next day said that they seemed to have somehow suffered a power "overload."

Once again a strange encounter involved pressure changes, spatial relocation, electrical effects, and a major temporal disruption. At face value these stories do not describe little green men, but some natural phenomenon that transported the occupants and their car into the future. I have now collected more than 300 such cases, involving a consistent array of effects on the environment that are amenable to scientific study.

I make absolutely no presumptions about these matters. It may be that independent exploration will remove any connection I feel they might have to time travel. However, my sense is that these could just be the missing clue that will make all things fall into place. We are too close to success to fail to investigate any realistic possibility because of cultural biases about what these events might represent. At the very least, I feel these cases are worthy of investigation.

REFERENCES

1 The first time that a Bose-Einstein condensate was actually created was in 1995, when rubidium gas was cooled to just ten billionths of a degree above absolute zero. It existed for under a minute before turning into an icy sludge when the temperature rose very slightly.

2 Anonymous, "Creation of a new state of matter," *New Scientist*, October 9, 2001.

3 Report on worldnetdaily.com, September 30, 2003.

4 Warner, Rex, trans., and St. Augustine. *The Confessions of St. Augustine*. New York American Library, 1963.

5 Caveney, Peter and Highfield, Roger *The Arrow of Time*. London: WH Allen, 1990.

6 Interviews with aboriginal villagers in the Kakadu region, Northern Territories, Australia, September 1991.

7 Lippincott, Kristen. *The Story of Time*. London: Merrell Holberton, 1999.

References

8 Randles, Jenny *Supernatural Isle of Man*. London: Robert Hale, 2003.

9 Whitrow, G. J. *The Natural Philosophy of Time*. Oxford: Oxford University Press, 1980.

10 Kepler, Johannes. *Somnium*, Private manuscript, 1634.

11 Irving, Washington. *Rip van Winkle*. New York: Gramercy, 1999 (originally published in 1850).

12 Twain, Mark. *A Connecticut Yankee in King Arthur's Court* in *Four Complete Novels*. New York: Gramercy, 1993 (originally published in 1889).

13 See letter from meteorologist Norman Brooks in *Fortean Times*, London, July 2001.

14 Wells, H. G. *The Time Machine*. New York: Tor Books, 1995 (originally published in 1895).

15 Nichelson, Oliver. "The Death Ray of Nikola Tesla," *Fate*, January 1990.

16 See www.alienshift.com

17 Moriny, Gary F. *The Complete Idiot's Guide to Understanding Einstein*. New York: Alpha Books, 2000.

18 Hey, Tony and Patrick Walles. *Einstein's Mirror*. Cambridge: Cambridge University Press, 1997.

19 Einstein, Albert. *Relativity: The Special and General Theory*. London: Methuen, 1920.

20 Eddington, Sir Arthur Stanley. *The Nature of the Physical World*. Cambridge: Cambridge University Press, 1928.

21 Hey, Tony and Tony Walles, op. cit.

22 Baird, John Logie. *Sermons, Soap and Television*. London: Royal Television Society, 1988.

23 Kaku, Michio. *Hyperspace: A Scientific Odyssey Through Parallel Universes, Time Warps and the Tenth Dimension*. Oxford: Oxford University Press, 1994.

24 Li, L. X. and J. R. Gott. "Self consistent quantum vacuum for Misner space allowing time travel," *Physical Review Letters*, 80, 1998.

25 Rae, Alistair I. M. *Quantum Physics*. Cambridge: Cambridge University Press, 1994.

26 Charroux, Robert. *One Hundred Thousand Years of Men's Unknown History*. Suffolk: Spearman, 1965.

27 Godel, K. "An example of a new type of cosmological solutions of Einstein's equations of gravitation," *Reviews of Modern Physics*, Vol. 21, No. 3, July 1949.

28 Hoyle, Fred. *The Black Cloud*. New York: Buccaneer, 1957. Hoyle, Fred and Chandra Wickramasinghe. *Lifecloud: Origin of Life in the Universe*. London: Dent, 1978.

29 Gott, J. Richard III. *Time Travel in Einstein's Universe: The Physical Possibilities of Travel Through Time*. New York: Houghton Mifflin, 2001.

30 Hoyle, Fred. *October the First Is Too Late*. London: Gollancz, 1966.

31 Tipler, Frank. "Causality violation in asymptotically flat spacetimes," *Physical Review Letters*, 37, pp. 879–82, 1976.

32 Gribbin, John. *Unveiling the Edge of Time: Black Holes, White Holes, Wormholes*. New York: Crown Publications, 1992.

33 Chambers, J. "Letter," *Fortean Times*, 165, December 2002.

34 Krause, Peter. *Father Ernetti's Chronovisor: The Creation and Disappearance of the World's First Time Machine*. Florida: New Paradigm Books, 2000.

35 Brune, François. *New Mysteries of the Vatican*. Editions Albin Michel. Paris: 2002.

36 Deutsch, David and Michael Lockwood. "The quantum physics of time travel," *Scientific American*, March 1994.

37 Gott, 2001, op. cit. pp. 12–18.

38 Gott, J. R. "A time-symmetric, matter and anti-matter tachyon cosmology," *Astrophysical Journal*, 187, 1974.

39 Benford, Gregory. *Timescape*. New York: Pocket Books, 1980.

40 Greene, op. cit., Chapter 7.

41 Feynman, Richard P. *The Character of Physical Law*. Cambridge: MIT Press, 1995.

42 Hawking, Stephen. *A Brief History of Time*. New York: Bantam, 1996.

43 See www.new-fiction.co.uk

44 Gott, J. R. "Will we travel back in time?" *Time*, April 10, 2000.

45 Moriny, op. cit.

46 *Science News Digest*, NASA, January 23, 2004.

47 Manoharan, H. C. Lutz, & D. Eigler. "Quantum mirages formed by coherent projection of electronic structure," *Nature*, Vol. 403, pp. 512–5, February 3, 2000.

48 Deutsch, David. *The Fabric of Reality*. London: Penguin, 1998.

49 Randles, Jenny. *Time Storms: The Amazing Evidence of Time Warps, Space Rifts and Time Travel*. London: Piatkus, 2001.

50 Semeniuk, I. "Are you fed up with absurd notions of quantum weirdness?" *New Scientist*, March 9, 2002.

51 Davies, Paul. "How to build a time machine," *Scientific American*, September 2002.

52 Sagan, Carl. *Contact*. New York: Simon & Schuster, 1985.

53 Morris, Michael S., Kip S. Thorne, and Ulvi Yurtsever. "Wormholes, time machines and the weak energy condition," *Physical Review Letters*, Vol. 26, No. 13, September 26, 1988.

54 Thorne, K. 1988, op. cit.

55 Hawking, Stephen. "Chronology protection conjecture," *Physical Review Letters D* 46, pp. 603–11, 1992.

56 Semeniuk, I. "No going back," *New Scientist*, September 20, 2003.

57 Dyson, Lisa. "Chronology protection in string theory," www.arxiv.org.

58 Konkowski, Deborah and William Hiscock, "Vacuum state for Misner space," *Physical Review D*, 26, 1982.

59 Novikov, Igor *The River of Time*. Cambridge: Cambridge University Press, 1998.

60 Thorne, Kip. *Black Holes and Time Warps*. New York: Norton, 1994.

61 An interview with Gibbs appears in *Strange Magazine*, 14, 1994.

62 A circuit diagram of Gibbs's device appears in *Strange Magazine*, 15, 1995.

63 Gott, J. R. "Gravitational lensing effect of vacuum strings: exact solution," *Astrophysical Journal*, 288, p. 422, 1985.

64 Persinger, Michael A. and Gylaine F. Lafreniere. *Space-Time Transients and Unusual Events.* New York: Nelson-Hall, 1977.

65 Greene, Brian. *The Elegant Universe: Superstrings, Hidden Dimensions, and the Quest for the Ultimate Theory.* London: Jonathan Cape, 1999.

66 Gott, J. R. "Closed time-like curves produced by pairs of moving cosmic strings: exact solutions," *Physical Review Letters,* 66, p. 1126, 1991.

67 Cutler, C. "Global structure of Gott's two string space-time," *Physical Review Letters D,* 45, pp. 487–94, 1992.

68 Based on notes taken at Imperial College, April 1976.

69 Cohen, D. "Anti-gravity research on the rise," *New Scientist,* July 30, 2002.

70 Hambling, D. "Contra anti-gravity," *Fortean Times,* January 2003.

71 Hoyle, Trevor. *Seeking the Mythical Future.* New York: Ace, 1982 (originally published in 1977).

72 Deutsch, D. and M. Lockwood, *Scientific American,* op. cit.

73 Report by the Quantum Imaging Lab, Boston University, November 2001.

74 Wheeler, John Archibald and Kenneth Ford. *Geons, Black Holes and Quantum Foam.* New York: Norton, 1999.

75 Clarke's research notes can be read at www.cix.co.uk/
-sjbradshaw/baxterium/light-outline.html
Baxter, Stephen and Arthur C. Clarke. *The Light of Other Days.* New York: St. Martin's Press, 2000.

76 Kaku, online forum, May 2004.

77 Flynn, M. "The Ultimate Speed Limit?" *X Factor,* 58, 1999.

78 Study by NEC lab, New Jersey, July 2000, and numerous subsequent media reports.

79 Chown, M. "Unwrite This," *New Scientist,* November 27, 1999.

80 Ibid.

81 Gell-Mann, Murray. *The Quark and the Jaguar.* New York: WH Freeman, 1994.

82 Gell-Mann, op. cit.

83 Price, H. *Quantum Handshakes*. Sydney: University of Sydney, 2002.

84 Mallett, Ronald. "Weak gravitational field of the EM radiation in a ring laser," *Physical Review Letters* 269, 2000.

85 Furusawa, A. "Unconditional quantum teleportation," *Science*, 282, October 23, 1998.

86 For details of the Innsbruck project see: Bouwmeester, D., et al. "Experimental quantum teleportation," *Nature* 390, pp. 575–9, December 11, 1997.

87 For further data on early teleports see: Bennett, C. "Teleporting an unknown quantum state," *Physical Review Letters* 70, 1993. For an update see: Barrett, M. "Deterministic quantum teleportation of atomic qubits," *Nature* 429, pp. 78–9, June 17, 2004.

88 Press release by the ANU—Australian National University, Canberra, Australia, June 17, 2002.

89 Crichton, op. cit. Filmed as *Timeline*, released Fall 2003.

90 Diagram of the prototype device available via Moscow Aviation Institute, Panfilov Str 20-2, Moscow, 125080 Russia.

91 Chernobrov, Vadim. "Experiments on the change of direction and rate of time motion," Proceedings of the international conference—new ideas in natural sciences—St. Petersburg, 1996.

92 Hendry, Allan. *The UFO Handbook: A Guide to Investigating, Evaluating, and Reporting UFO Sightings*. London: New English Library, 1981.

Case files of the Center for UFO Studies, Evanston, Illinois.

93 Randles, Jenny. *Time Storms: The Amazing Evidence of Time Warps, Space Rifts and Time Travel*. London: Piatkus Books, 2000.

INDEX

Index

Index

DANVILLE PUBLIC LIBRARY
DANVILLE, INDIANA